Peripheral Territories, Tourism, and Regional Development

Edited by Rui Alexandre Castanho, Gualter Couto and Rossana Santos

Published in London, United Kingdom

IntechOpen

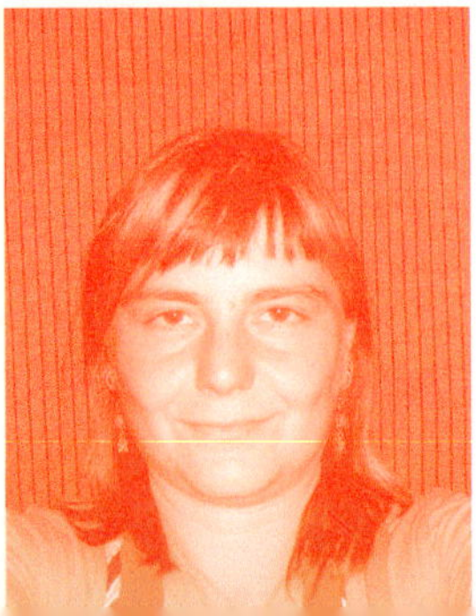

Supporting open minds since 2005

Peripheral Territories, Tourism, and Regional Development
http://dx.doi.org/10.5772/intechopen.93284
Edited by Rui Alexandre Castanho, Gualter Couto and Rossana Santos

Contributors
Joaquim Mourato, Ana José, Deolésio Mendes, Daniel F. Meyer, Ana Vulevic, Rui Alexandre Castanho, José Manuel Naranjo Gómez, José Cabezas Fernández, Luis Fernández-Pozo, Jacinto Jacinto Garrido Velarde, José Martín Gallardo, Luis Loures, Sérgio António Neves Lousada, Pedro Pimentel, André Oliveira, Gualter Manuel Medeiros Do Couto, João Crispim Ponte, José Manuel Jurado Almonte, Julian Mora Aliseda, Luiz Pinto Machado, António Almeida, Rossana Santos

First published in London, United Kingdom, 2021 by IntechOpen
IntechOpen is the global imprint of INTECHOPEN LIMITED, registered in England and Wales, registration number: 11086078, 5 Princes Gate Court, London, SW7 2QJ, United Kingdom
Printed in Croatia

British Library Cataloguing-in-Publication Data
A catalogue record for this book is available from the British Library

Additional hard and PDF copies can be obtained from orders@intechopen.com

Peripheral Territories, Tourism, and Regional Development
Edited by Rui Alexandre Castanho, Gualter Couto and Rossana Santos
p. cm.
Print ISBN 978-1-83968-183-7
Online ISBN 978-1-83968-184-4
eBook (PDF) ISBN 978-1-83968-443-2

For EU product safety concerns:
IN TECH d.o.o., Prolaz Marije Krucifikse Kozulić 3, 51000 Rijeka, Croatia, info@intechopen.com or visit our website at intechopen.com.

Meet the editors

Rui Alexandre Castanho has a master's degree in Planning, Audit, and Control in Urban Green Spaces and an international Ph.D. in Sustainable Planning in Borderlands. Currently, he is a professor at WSB University, Poland, and Coordinator of International Relations at the Research Institute on Territorial Cooperation and Inter-Organizational Cooperation. He is also a visiting professor at the University of Johannesburg, South Africa. Dr. Castanho is a post-doc researcher on the GREAT Project, University of Azores, Ponta Delgada, Portugal. He collaborates with the Environmental Resources Analysis Research Group (ARAM), University of Extremadura (UEx); VALORIZA - Research Center for the Enhancement of Endogenous Resources, Polytechnic Institute of Portalegre (IPP), Portugal; CITUR - Madeira - Centre for Tourism Research, Development and Innovation, Madeira, Portugal; and AQUAGEO Research Group, University of Campinas (UNICAMP), Brazil.

Gualter Couto obtained a BSc in Business Management from Universidade dos Açores, Portugal, in 1995; an MSc in Business Management and an MBA from Instituto Superior de Economia e Gestão da Universidade de Lisboa, in 1999; and a Ph.D. in Management/Finance from Instituto Superior de Economia e Gestão da Universidade de Lisboa, in 2006. Since 1996, Dr. Couto has been a Finance Professor at the School of Business and Economics, Universidade dos Açores. Since 2003, he has been a researcher at the Center of Applied Economics Studies of the Atlantic. He is an expert in corporate finance and previously served as director of the MBA program at Universidade dos Açores.

Rossana Santos is a full-time teacher and current coordinator of the BA in Hotel Management and Administration in the School of Technologies and Management, University of Madeira, Portugal. She holds a Ph.D. in Tourism (2013), a master's degree in Tourism Management and Development (2004), and a BA in Tourism (2001). Since 2006, she has been teaching in tourism and hospitality courses at several Portuguese institutions. She has also published several scientific papers on tourism.

Contents

Preface

Limited land and resources, along with the overexploitation of tourism and multiple other factors, make peripheral and ultra-peripheral territories relevant cases for studying governance and sustainable development.

In the specific case of island regions (ultra-peripheral territories), tourism regularly assumes itself as one of the main, if not the most critical, regional economic drivers. Alternatively, in peripheral territories like the borderlands, it is evident that these regional synergies have reached unprecedented levels in the last few years, not only due to their potential for territorial integration but also recognizing their role in supranational processes (i.e., infrastructure construction and planning activities on European territories). Cross-Border Cooperation (CBC) practices have increased in Europe and worldwide, generating a global network of relationships between people to create mutually beneficial scenarios.

Contextually, the study of these typologies of territories—islands and borderlands—is essential to plan the future of our regions, businesses, and societies. Therefore, it is critical to analyze the territorial subjacent processes behavior based on the new challenges as an example of sustainable development and growth.

This book includes contributions from renowned academicians and technicians working in the field of sustainable tourism in rural and peripheral areas. The content contained herein provides the reader with a deep understanding, from several perspectives, of the dynamics, challenges, and opportunities of tourism in these specific territories.

Peripheral Territories, Tourism, and Regional Development is a significant scientific contribution to this specific field.

Rui Alexandre Castanho
WSB University,
Dabrowa Górnicza, Poland

Rossana Santos
University of Madeira,
Funchal, Portugal

Gualter Couto
University of Azores,
Ponta Delgada, Portugal

Section 1

Peripheral Territories, Tourism, and Regional Development: A Brief Overview

Chapter 1

Introductory Chapter: Rural Tourism as a Catalyst for Sustainable Regional Development of Peripheral Territories

Rui Alexandre Castanho, Gualter Couto and Rossana Santos

1. Introduction

1.1 Sustainable development and regional development: a brief overview

Sustainable Development is no longer a choice, but a necessity of us all. In fact, if we look to prosper as a society, and probably as a specie, there is no alternative instead of opting for a typology based on a sustainable development and growth.

This development is often defined as the process of meeting the demands of today without jeopardizing the necessities of future generations without limiting their possibilities to plan the territory in their own way [1–7].

The term 'sustainable development' was firstly introduced in the international policy debate by the World Strategy for Nature. It was established as a new world paradigm after 'Our Common Future,' the final report of the Brundtland Commission [8–10].

The starting point of the enlarged use of the sustainable development concept was the United Nations conferences on Environment and Development [8]. In those conferences, the necessity to adopt new growth and development strategies as locally as globally has been admitted. This demand was previously discussed in 1969 through a study conducted by UNESCO, which revealed that, by 2000, the urban population in rural areas throughout the world would be solely 15%. In 1993, the World Bank stated that 2010 is the year in which most of the world's population would live in cities, creating a dramatic scenario for urban systems, desertification dilemmas, and the rarity of resources in rural areas [2–5].

Accordingly, sustainable development is combined and joined with sustained economic growth, social equity, and inclusion. Besides, it is also linked to environmental preservation and conservation, requiring a robust political commitment, local and regional stakeholders' involvement, and public participation [11, 12]. Hence, from the several instruments used by sustainable development, territorial planning is one of the most relevant for this development typology as in urban as in rural areas. Through this development is expected to strengthen regional and local economies, promote social development and cohesion, conceive a conscience more directed for the environmental issues, and create inclusive and safer territories [13, 14].

The concept of sustainable development is broad and comprehensive and is based on three fundamental pillars economic, social, and environmental. In fact, the effects of these pillars are very dynamic and multifaceted [7, 15]. In addition to its versatility, sustainable development's progressive nature makes it challenging to comprehend and evaluate the problem in full. Hence, sustainable development should be measured with sufficient economic, social, and environmental indicators and indices applying suitable techniques [15, 16].

While sustainable development indicators provide clear and straightforward information to the public and decision-makers on the subject, indices are seen as a tool that summarizes the complex realities of sustainable development, ensure accountability of decision-makers against the results of their policies and enable the public to understand sustainable development better [17].

Thereby, we should look more closely at the three main pillars of sustainable development: economic, social, and, environmental. According to Spangenberg [16], the economic pillar could be understood as "(...) *a particular subsystem of society, which due to its features such as the specific logic of efficiency along with the short-term time frames, permits us to understand human beings as a profit-maximizing individual*". In fact, sustainability claims for an economic system that matches the necessities of its populations, offering enough employment and rejuvenating its population to address these services in the long-term [16]. So, to meet these requirements, the economic system's competitiveness must be an essential part of the concept of sustainability [16]. Moreover, it can be extrapolated away from this description and include regional and local economic development models, land-uses and land covers, real estate market, among numerous others [18–21].

The social sphere, usually, refers to public policies that support social challenges. Social problems are related to our common well-being and prosperity, including elements as healthcare, education, housing, or employment, are just a few examples [7]. They ensure that individuals do have access to social services; besides, they do not be affected by the lack of knowledge of their rights and practice a responsible contribution to social services and policies, as on a local scale as in a national [22]. Furthermore, many authors defend that also, the institutional dimension should be strongly considered. Jörg Spangenberg [15] states that: "*Institutions are the success of the social interactions, along with established rules over the society, by the decision-making processes and their tools to apply such policies. So, the institutional dimension includes groups from civil society and the policy-makers, from the administrative system, and technicians*". Consequently, if we look through a Sustainable Development attitude, it becomes clear the relevance of the public participation, the equality opportunities, or the no social discrimination, along with the strong political responsibility and transparency [7, 23].

The environmental pillar is described as the aggregate of all the bio-geological processes along with their constituents. Accordingly, it requires the preservation and conservation of the ecological systems as a natural basis to support the Anthropological sphere [15, 24]. Consequently, through well-designed and executed planning strategies, the combination and interaction within societies and the environment may produce diverse advantages for cities and territories in different contexts. Also, before-mentioned synergies sustain green areas with ecological and cultural heritage values, as is the preservation of biodiversity, prevention of the generation of heat islands in urban areas, amid various additional benefits [25–27].

Besides, we should recognize the incongruities resulting from the disparities in the planning goals. In this regard, we may name some as the different interests of local and regional stakeholders, the problems related to waste management planning, bureaucracies' and disparities regarding land-uses, are just some of the challenges that occur in rural and urban territories.

Such planning problems shall be used as agitators to foster economic performance, enhance social equity, and boost environmental efficiency instead of being sustainable development barriers.

Although decisive and confident government policies that set people in the leading role are required to bring development to the vanguard, furthermore, sustainable development is essential for the continuity of the environment, which is increasing the capability to use the ecological environment and natural resources and distribute the resources equitably among individuals [17].

2. Rural tourism and peripheral territories

The rural world has been and continues to face several crises, increase in unemployment, aging of the population, emigration, negative migratory balances, and accelerated restructuring of production and farms. In these peripheral territories, economic development plays an essential role in triggering sustainable development. For this reason, the potential of rural tourism, linked with entrepreneurship and small and medium-sized enterprises (SMEs), is emphasized by authors as they comprise the volume of local business activity [28].

This argument is based on evidence of previous studies related to the impact of economic activities on employment and income in peripheral territories over time. In the period that followed World War II, the agricultural sector's contribution to the Gross National Product (GNP) and employment source had been decreasing [29]. As a result, the strategies addressed for rural development began to be concentrated on industrial activities.

However, evidence concerning major industry location preferences in urban centers [30], as well as larger indirect multiplier effects of employment created by big industries in some developing countries, subsists. Meller and Marfán [31] suggest that the industrial sector is not the most suitable strategy to trigger sustainable development processes in all contexts of peripheral territories. On the other hand, when considering the construction economic activity sector, King [32] states that jobs created are temporary as their continuity depends on the regular flow of emigrants returning to the areas of origin.

Furthermore, during the 80s, the increase in employment in peripheral territories of the Organization for Economic Cooperation and Development was mainly due to growth in the services sector [33–34]. In this sense, as argued by King [35], the increase in employment in the tertiary sector has proved to be the key variable in explaining immigration or low emigration in those territories.

Still, not all tertiary sector economic activities show the same ability to generate employment and income, acknowledging their decreasing importance as employers during the 90s [33]. Alternatively, the role of tourism has made a major contribution to sustainable development [29]. Several studies conducted in these territories have evidenced that tourist spending creates more jobs and income than any other sector of the economy, and it generates and maintains employment in other sectors of the economy that support or provide visitors and tourism companies [36–67].

UNWTO [68] understands Rural Tourism as "*a type of tourism activity in which the visitor's experience is related to a wide range of products generally linked to nature-based activities, agriculture, rural lifestyle/culture, angling and sightseeing*". *In this scope, "Rural Tourism activities take place in non-urban (rural) areas with the following characteristics: i) low population density, ii) landscape and landuse dominated by agriculture and forestry and iii) traditional social structure and lifestyle"*.

In the context of the COVID-19 pandemic, the UNWTO [68] also states that rural tourism is even more relevant because tourists look for uncongested destinations, preferably with open-air experiences and activities.

3. Closing remarks

Tourism development in peripheral rural areas has had little attention, and most rural communities are not prepared to deal with structural changes in their economy. People who live in these communities commonly work in agriculture or fisheries, have low education, and do not have the appropriate skills to establish synergetic relationships with tourists and tourism companies. So, it is critical to capitalize on the strengths that exist in these communities, especially regarding their bond with natural resources, traditional economic activities, folklore, food, festivities, and rituals. Through these features, we need to try to identify ways to structure innovative tourism experiences in rural areas and add value to local communities, contributing to reducing poverty and social exclusion and the increase in employment.

Based on the above-mentioned, it is possible to realize the relevance of following a sustainable development process to design and implement the upcoming territorial plans and policies. Contextually, it is disclosed how rural and peripheral areas, through rural tourism activities, could add to the regional sustainable development achievement [69, 70].

Furthermore, we should also consider that the development of sustainability considering tourism is an ongoing process that requires constant monitoring of impacts, introducing preventive and/or corrective measures when necessary. Consequently, all the relevant involved actors' active participation is essential, simultaneously with strong political leadership – to build consensus and engage other actors in this sustainability reaching process [68–70].

Funding

This paper is financed by Portuguese national funds through FCT – Fundação para a Ciência e a Tecnologia, I.P., project number UIDB/ECO/00685/2020 and also by the project GREAT - Genuine Rural Experiences in the Azores Tourism with the code: ACORES-01-0145-FEDER-000089.

Author details

Rui Alexandre Castanho[1,2,3,4,5]*, Gualter Couto[1] and Rossana Santos[4,6]

1 School of Business and Economics and CEEAplA, University of Azores, Ponta Delgada, Portugal

2 Faculty of Applied Sciences, WSB University in Dąbrowa Górnicza, Poland

3 VALORIZA - Research Centre for Endogenous Resource Valorization, Portalegre, Portugal

4 CITUR - Madeira - Centre for Tourism Research, Development and Innovation, Funchal-Madeira, Portugal

5 CNPQ Research Group Aquageo Ambiente Legal, University of Campinas (UNICAMP), Campinas, Brazil

6 University of Madeira, Funchal-Madeira, Portugal

*Address all correspondence to: acastanho@wsb.edu.pl

References

[1] Loures, L. and Burley, J., (2012). Post-industrial land transformation – An approach to sociocultural aspects as catalysts for urban redevelopment. In: Burian, J. (Ed.) Advances in Spatial Planning, IntechOpen, pp. 223-246. (DOI: 10.5772/2123) (ISBN: 978-953-51-0377-6).

[2] Loures, L., (2011). Planning and Design in Post-industrial Land Transformation: East bank Arade river, Lagoa - case study, Faculdade de Ciências e Tecnologia, Universidade do Algarve, Faro, Portugal – Doctoral Thesis

[3] Castanho, R. A (2017). Sustainable Urban Planning in Transboundary Areas: Analysis of Critical Factors for Territorial Success. Doctoral Thesis University of Extremadura (UEx) Badajoz, Spain. Department of Vegetal Biology, Ecology and Earth Sciences.

[4] Loures L., Santos, R. and Panagopoulos T., (2007). Urban Parks and Sustainable Development – The Case Study of Portimão City, Portugal. In: Stamou, A., Panagopoulos, T., Stamatiou, E., Hatzopoulou, A. e Antunes, D. (Eds.) Energy, Environment, Ecosystems & Sustainable Development, Athens: WSEAS Press, pp.127-131. (ISBN: 978-960-8457-88-1).

[5] Castanho, R.A., Cabezas, J., Loures, L., and Fernández-Pozo, L. (2017). Evolution of Urban Planning Process in the Iberian Peninsula and his Footprints in Cityscape. New Challenges. Monfragüe Resiliente – Scientific Journal. Vol. VIII (2). ISSN: 2340-5457.

[6] Ulucak, R., Yücel, A. and Koçak, E. (2019). The process of sustainability: From past to present. Environmental Kuznets Curve (EKC)

[7] Castanho, R.A., Couto, G., & Pimentel (2020). Principles of Sustainable Tourism and Cultural Management in Rural and Ultra-peripheral Territories: Extracting Guidelines for its application in the Azores Archipelago. Cultural Management: Science and Education (CMSE) ISSN: 2512-6962 DOI: 10.30819/cmse.4-1.01

[8] Spangenberg, J. (1995), Towards Sustainable Europe, A Study from the Wuppertal Institute for Friends of the Earth Europe, Luton, UK.

[9] Spangenberg, J. (2000), Sustainable Development – Concepts and Indicators, Colonia, Germany.

[10] Castanho, R.A., Kurowska-Pysz, J. and Naranjo Gómez, J. (2019). How to Reach the Eurocities? A Retrospective Review of the Evolution Dynamics of Urban Planning and Management on the Iberian Peninsula Territories. Sustainability 2019, Special Issue Sustainable Cross-Border Cooperation: Common Planning, Policies, Strategies, Methods and Activities. Sustainability 2019, 11, 602; pp. 6-23. doi:10.3390/su11030602.

[11] Faludi, A., & Valk, A. (1994). Rule and Order Dutch Planning Doctrine in the Twentieth Century. GeoJournal Library ; v. 28. https://doi.org/10.1007/978-94-017-2927-7

[12] Loures, L., Naranjo Gómez, J.M., Castanho, R.A., Loures, A., (2020). Benefits and Limitations of Public Involvement Processes in Landscape Redevelopment Projects—Learning from Practice. Chapter in the Book: Regional Intelligence Spatial Analysis and Anthropogenic Regional Challenges in the Digital Age. Ed. Springer. pp. 29-48. DOI: 10.1007/978-3-030-36479-3_3.

[13] Amado, M. (2009). Planeamento Urbano Sustentável. Lisboa, Portugal.

[14] Fadigas, L., (2015): Urbanismo e TerritoÃÂrio - As politicas publicas. Lisboa, Portugal.

[15] Spangenberg, J., (2000). Sustainable Development – Concepts and Indicators. Colonia, Germany.

[16] Spangenberg, J., (1995). Towards Sustainable Europe, A Study from the Wuppertal Institute for Friends of the Earth Europe. Luton, UK.

[17] United Nations (2020). Promote peaceful and inclusive societies for sustainable development, provide access to justice for all and build effective, accountable and inclusive institutions at all levels. Available at: https://sdgs.un.org/goals/goal16 (accessed on August 26, 2020)

[18] Naranjo Gómez, J.M., Loures, L., Castanho, R.A., Cabezas, J., Fernández-Pozo, L., Lousada, S., Escórcio, P. (2018). Assessing Land Use Changes in European Territories: A Retrospective Study from 1990 to 2012 [Working Title], IntechOpen, DOI: 10.5772/intechopen.80827.

[19] Codosero Rodas, J., Cabezas, J., Naranjo Gómez, J., and Castanho, R.A., (2019). Risk Premium Assessment for the Sustainable Valuation of Urban Development Land: Evidence from Spain. Sustainability 2019, 11, 4191; pp-1-20 doi:10.3390/su11154191

[20] Codosero Rodas, J., Castanho, R.A., Cabezas, J., and Naranjo Gómez, J. (2020). Sustainable valuation of land for development. Adding value with urban planning progress. A Spanish case study. Land Use Policy 92 (2020) 104456 https://doi.org/10.1016/j.landusepol.2019.104456.

[21] Couto, G., Pimentel, P., Ponte, J. C., Gonçalves, P., Duarte, D., Arruda, A., . . . Rodrigues, F. (2018). Azores: Investor's guide to sustainable tourism. Ponta Delgada: FLAD - Fundação Luso Americana para o Desenvolvimento.

[22] European Union (EU), (2020). Policy Forum on Development. Working Group: Social Pillar of Sustainable Development. www. europa.eu (accessed on August 26, 2020).

[23] Castanho, R.A. (2020). The Relevance of Political Engagement and Transparency in Cross-Border Cooperation (CBC) Environments. Analyzing Border cities in Europe. Lex localis - Journal of Local Self-Government, Vol. 18, No. 3, pp. 487-502, July 2020. https://doi.org/10.4335/18.3.487-502(2020).

[24] Yigitcanlar, T., Dur, F., Dizdarogluc, D., (2015). Towards prosperous sustainable cities: A multiscalar urban sustainability assessment approach, Habitat International: a journal for the study of human settlements p36-46.

[25] Raposo, M., Castanho, R.A., Machado, M., Castro, C., Santos, P. and Pinto-Gomes, C. (2018). The Relevance of Vegetation Series on the Maintenance and Sustainability of Public Spaces in the South-west Iberian Peninsula. Chapter in the Book: Landscape Architecture: The Sense of Places, Models and Applications. Ed. InTech. ISBN 978-1-78923-713-9. pp.254-272.

[26] Castanho, R.A. Cabezas, J., Naranjo Gómez, J., Martín Gallardo, J., Fernández-Pozo, L., Loures, L., Yilmaz Genç, S., and Lousada, S., (2020). Assessing Ecosystem Services delivered by Public Green Spaces in European Major Cities. Chapter in the Book: Landscape Architecture. Ed. Intech. DOI: http://dx.doi.org/10.5772/intechopen.91415

[27] Kaletová, T., Loures, L., Castanho, R.A., Aydin, E., Gama, J., Loures, A., Truchy, A. (2019). Relevance of Intermittent Rivers and Streams in Agricultural Landscape and Their

Impact on Provided Ecosystem Services—A Mediterranean Case Study. pp. 3-16. Int. J. Environ. Res. Public Health (IJEPRH) 2019, 16, 2693; doi:10.3390/ijerph16152693

[28] Labrianidis, L.; Ferrão, J.; Hertzina, K.; Kalantaridis, C.; Piasecki, B.; Smallbone, D. The Future of Europe's Rural Periphery; Final Report; 5th Framework Programme of the European Community; Routledge: Thames, UK, 2003.

[29] Williams, A. Introduction. In Southern Europe Transformed-Political and Economic Change in Greece, Italy, Portugal and Spain;Williams, A., Ed.; Harper and Row: New York, NY, USA, 1984; pp. 1-32.

[30] Hayter, R. The Dynamics of Industrial Location—The Factory, the Firm and the Production System; John Wiley and Sons: London, UK, 1997.

[31] Meller, P.; Marfán, M. Small and Large Industry: Employment generation, linkages, and key sectors. Econ. Dev. Cult. Chang. 1981, 29, 263-274.

[32] King, R. Return Migration and Regional Economic Development. In Return Migration and Regional Economic Problems; Croom Helm: Sydney, Australia, 1986; pp. 1-37.

[33] Bryden, J.; Bollman, R. Rural Employment in industrialized countries. Agric. Econ. 2000, 22, 185-197. [CrossRef]

[34] Kilkenny, M.; Partridge, M. Export sectors and rural development. Am. J. Agric. Econ. 2009, 91, 910-929. [CrossRef]

[35] King, R. Population Mobility: Emigration, Return Migration and Internal Migration. In Southern Europe Transformed—Political and Economic Change in Greece, Italy, Portugal and Spain;Williams, A., Ed.; Harper and Row: New York, NY, USA, 1984; pp. 145-178.

[36] Bergstrom, J.; Cordell, K.; Ashley, G.; Watson, A. Economic impact of recreational spending on rural areas: A case study. Econ. Dev. Q. 1990, 4, 29-39. [CrossRef]

[37] Fleischer, A.; Felsenstein, D. Support for rural tourism—Does it make a diference? Ann. Tour. Res. 2000, 27, 1007-1024. [CrossRef]

[38] Balaguer, J.; Jordá, M. Tourism as a long-run economic growth factor: The Spanish case. Appl. Econ. 2010, 34, 877-884. [CrossRef]

[39] Mahony, K.; Zyl, J. The Impacts of Tourism Investment on Rural Communities: Three Case Studies in South Africa. Dev. S. Afr. 2002, 19, 83-103. [CrossRef]

[40] Vanegas, M.; Croes, R. Growth, development and tourism in a small economy: Evidence from Aruba. Int. J. Tour. Res. 2003, 5, 315-330.

[41] Hughes, D.; Shields, M. Revisiting tourism regional economic impact: Accounting for secondary household employment. Rev. Reg. Stud. 2007, 37, 186-206.

[42] Mazumder, H.; Ahmed, M.; Al-Amin, Q. Estimating total contribution of tourism to Malaysia economy. Int. J. Bus. Manag. Soc. Sci. 2009, 2, 146-159.

[43] Chen, K.; Yang, H. Appraising the economic impact of the opening up to mainland Chinese tourist arrivals' policy on Taiwan with a tourism—CGE model. Asia Pac. J. Tour. Res. 2010, 15, 155-175. [CrossRef]

[44] Haddad, E.; Porsse, A.; Rabahy, W. Domestic tourism and regional

inequality in Brazil. Tour. Econ. 2013, 19, 173-186. [CrossRef]

[45] Klytchnikova, I.; Dorosh, P. Tourism sector in Panama: Regional economic impacts and the potential to benefit the poor. Nat. Resour. Forum 2014, 32, 70-79.

[46] C.; Lee, M.; Yoon, S.H. Estimating the economic impact of convention and exhibition businesses, using a regional input-output model: A case study of the Daejeon Convention Center in South Korea. Asia Pac. J. Tour. Res. 2013, 18, 330-353. [CrossRef]

[47] Snyman, S. Household spending patterns and flow of ecotourism income into communities around Liwonden National Park, Malawi. Dev. S. Afr. 2013, 30, 640-658. [CrossRef]

[48] Fodranov, I.; Kubickova, V.; Michalkova, A. Measuring societal value of tourism: A new approach. Tourism 2015, 63, 423-434.

[49] Hyojin, K.; Byung-Gook, K. Economic impacts of the hotel industry: An input-output analysis. Tour. Rev. 2015, 70, 132-149.

[50] Stephen, P. Potential Economic contribution of regional tourism development in China: A comparative analysis. Int. J. Tour. Res. 2015, 17, 303-312.

[51] Crompton, J.; Ji, Y.; Dudensing, R. Sources of variation in economic impact multipliers. J. Travel Res. 2016, 55, 1051-1064. [CrossRef]

[52] Yesavdar, U.; Belgibayav, A.; Mersakylova, G. The role of developing direction of international tourism in Kazakhstan. Bull. Natl. Acad. Sci. Repub. Kazakhstan 2016, 2, 180-185.

[53] Croes, R.; Rivera, M. Tourism's potential to benefit the poor: A social accounting matrix model applied to Ecuador. Tour. Econ. 2017, 23, 29-48. [CrossRef]

[54] Kim, S.; Miller, C. An economic model comparison of EMSI and IMPLAN: Case of mistletoe marketplace. Tour. Econ. 2017, 23, 1124-1130. [CrossRef]

[55] Ruzic, P.; Demonja, D. Transformations in business & economics. Econ. Impacts Rural Tour. Rural Areas Istria 2017, 16, 31-40.

[56] Sánchez, V.; Fernandez, M.; Lara, J. Economic impact of a religious and tourist event: A holy week celebration. Tour. Econ. 2017, 23, 1255-1274. [CrossRef]

[57] Veretekhina, S.; Shinkareva, O.; Kozhaev, J.; Telepchenkova, N.; Kuznetsova, E.; Plekhanov, N. Evaluation methodology of the multiplier e ect for the Region as the result of the cluster formation. Eurasian J. Anal. Chem. 2017, 12, 533-547. [CrossRef]

[58] Yashalova, N.; Akimova, M.; Aleksandrovich, R. Propects for regional development of industrial tourism in view of the analysis of the main economic indicators of Russian tourism industry. Econ. Soc. Chang. Facts Trends 2017, 10, 195-213.

[59] Ferrari, G.; Mondejar, J.; Secondi, L. Tourists' expenditure in Tuscany and its impact on the regional economic system. J. Clean. Prod. 2018, 171, 1437-1446. [CrossRef]

[60] Hjerpe, E. Outdoor recreation as a sustainable export industry: A case study of the boundary waters wilderness. Ecol. Econ. 2018, 146, 60-68. [CrossRef]

[61] Pascariu, G.; Ibanescu, B. Determinants and implications of the tourism multiplier e ect in EU economies. Towards a core-periphery

pattern? Amfiteatru Econ. 2018, 20, 982-997. [CrossRef]

[62] Suresh, K.; Tiwari, A.; Uddin, G.; Ahmed, A. Tourism, trade, and economic growth in India: A frequency-domain analysis of causality. Anatolia Int. J. Tour. Hosp. Res. 2018, 29, 319-325. [CrossRef]

[63] Tohmo, T. The economic impact of tourism in Central Finland: A regional input-output study. Tour. Rev. 2018, 73, 521-547. [CrossRef]

[64] Cannonier, C.; Burke, M. The economic growth impact of tourism in small islands developing states—Evidence from the Caribbean. Tour. Econ. 2019, 25, 85-108. [CrossRef]

[65] Hasanlj, Y.; Baizakov, S.; Salihova, S. Assessment of the impact of tourism sector on the economy of Azerbaijan, Kazakhstan and Iran using input-output models. In Proceedings of the International Scientific Conference on Economic and Social Development—Socio Economic Problems of Sustainable Development, Baku, Azerbaijan, 14-15 February 2019.

[66] Hsu, P. Economic impact of wetland ecotourism: An empirical study of Taiwan's Cigu Lagoon area. Tour. Manag. Perspect. 2019, 29, 31-40. [CrossRef]

[67] Laterra, P.; Nahuelhual, L.; Gluch, M.; Sirima, X.; Bravo, G.; Monjeau, A. How are jobs and ecosystem services linked at the local scale? Ecosyst. Serv. 2019, 35, 207-218. [CrossRef]

[68] UNWTO. UNWTO Recommendations on Tourism and Rural Development. UNWTO. Madrid. 2020.

[69] Santos, R., Castanho, R.A. and Lousada, S., (2020). The Portuguese Emigrants' Return and the Impacts over Tourism Development in Rural Areas: Directions for a Sustainable Planning. Chapter 5 (pp.85-100). in the Book: Espacios y Sociedades en Transformación. Ed. Thomson Reuters Aranzadi – Spain. ISBN 978-84-1346-693-4.

[70] Santos, R., Castanho, R.A. and Lousada, S. (2019). Return Migration and Tourism Sustainability in Portugal: Extracting Opportunities for Sustainable Common Planning in Southern Europe. Sustainability 2019, 11, 6468; doi:10.3390/su11226468.

Section 2

Peripheral Territories: The Borderlands

Chapter 2

The Most Meridional Border in Europe. Demographic and Environmental Changes

José Manuel Naranjo Gómez, Jacinto Garrido Velarde, José Martín Gallardo, José Manuel Jurado Almonte, Julian Mora Aliseda and José Cabezas Fernández

Abstract

Between 38°50′30"N/7° 04'35 W and 37°10′13"N/7°23′38"W has located the southernmost border of Europe, which stands apart from Portugal and Spain, which separates part of Portugal and Spain, making it one of the furthest from the central European area. A feature of this Spanish-Portuguese border stretch is that it is closely linked to the Guadiana River, one of the major waterways that cross the Iberian Peninsula from east to west. In 1998, the Albufeira Convention was signed, promoting Iberian cooperation at a scientific and technical level to strengthen the links of communication and collaboration at the technical level - the challenge of shared management of the waters and international basins that affect both countries. The Convention presented challenges and objectives on which the competent administrations have been working since then. Another peculiarity of this territory, except in a few situations, is its low population density. Therefore, knowing the demographic and environmental changes of the municipalities bordering the 'Raya' (common name as the border is known in these areas) constitutes valuable information that leads to the territorial management of these peripheral areas.

Keywords: cross-border, low population density, sustainability, landscape, development, tourism

1. Introduction

In a world that is increasingly global and dwarfed both in distance and socioeconomically, it is the cultural and economic interactions of the border territories that ensure the relations, continuity, and prosperity of its inhabitants [1]. The way of interpreting borders has evolved following the processes of geopolitical events that questioned them because alterations were caused in the context in which they were created and subsequently existed [2].

Research in various disciplines, from anthropology to geography, through economics, history, and even environmental matters, has revealed the constant flows that occur in border areas [3–5]. These movements are especially more palpable in third world countries, where the central power is less evident and it is the local powers that take the initiative and interact on both sides of the border [6]. In any case, the generation of a border between two territories leads to a mismatch in many aspects such as political, cultural, linguistic, legal, economic, and ethnic, among others, as these territories remain under different orders [7].

This is the case of the line of separation between Spain and Portugal, the Spanish-Portuguese border, known as "La Raya or A Raia". It extends over 1214 km, being the longest of the European Union countries, and the oldest and most stable since it has undergone very few variations over the centuries of existence since the already distant treaties of Zamora (1143) and Alcañices (1297).

In its southernmost part, from the beginning of contact between the two countries with the Guadiana River (38° 50′30 "N / 7° 04′35), it begins to delimit a fluvial borderline, serving as such for the most of this lower section of the river, until it empties at 37° 10′13 "N / 7° 23′38" W. This portion of the border stands as the southernmost in Europe, which makes it one of the furthest from the central European area. A characteristic of this border section is that it is closely linked to the Guadiana River, one of the great watercourses that cross the Iberian Peninsula from east to west, serving as a physical separation between the populations on both sides, but also acting as a link between them for their peculiarities. In these territories, there are more pronounced linguistic contrasts than those detected further north, in the demarcation area between Portugal and the Spanish region of Galicia [8]. These differences have been losing strength as globalization, and especially European integration (1986), is leaving behind the old perception of the border as a wall, replacing it with a more current one in which they are seen as bridges that contribute to the cooperation, integration, and development [9–11].

The municipalities that border this border present eminently rural characteristics and a great demographic weakness that bears little relation to the current economic structure, which complicates their development [12]. The abandonment of the primary sector is a common process linked to industrialization, which is intensified in mountain areas, in areas settled on poor lands, and, in general, in border areas far from the decision-making and development centers of countries, currently constituting one of the main causes of landscape changes at the local scale [13]. These processes of abandonment of agricultural and livestock activities are spread worldwide [14, 15], appearing frequently in the Mediterranean basin during the last decades [16], linked to the decrease in population and, above all, in the labor force. Rural [17]. These series of circumstances often provoke complex phenomena that modify the regime in which disturbances in the landscape occur and are often overlooked in comparison with climate change and the fragmentation of the territory [18]. Among these disturbances we can consider the fires that, as has been observed in some abandoned territories, lead to increases in their frequency [16] and intensity, reaching in some areas up to more than twelve times the number registered before abandonment [19].

Even so, the alterations that abandonment is causing in ecosystems are not clear, in some cases it is causing a beneficial effect on diversity, but in other circumstances, this effect is the opposite, these changes being reflected in the heterogeneity of landscapes [20–24].

From both the continental and national or regional-local scales, the information obtained from land cover-uses and the evolution they suffer over the years is of primary interest to assess the impacts of anthropic pressures and incorporate them. in the organization and formulation of territorial policies. In this sense, and of great utility for this type of study, the CORINE Land Cover (CLC) is available, which is made up of a database of vector maps of the European territory, which it classifies into a series of homogeneous landscape units [25]. With a minimum mapping unit of 25 hectares and with a series of updates, four after the initial one, it stands as a good study instrument for the evolution of landscapes at the different levels already indicated above from the supranational [26, 27] to the local [28, 29], passing through the national [30, 31] in very diverse aspects such as environmental, agricultural livestock, spatial planning or socioeconomic.

Border areas characterized by low population density, as is our field of study, face future challenges in which transformation and sustainability must be harmonized, allowing opportunities to be found by identifying changes and adaptations in the medium-long term together with balanced conservation [32, 33].

In 1998, the Albufeira Agreement was signed, promoting Iberian cooperation at the scientific and technical level to strengthen communication and collaboration ties in the challenge of shared management of waters and international basins that affect both countries. This agreement posed challenges and objectives on which the competent administrations have been working since then. Another peculiarity of this territory, except in a few situations, is its low population density, which has allowed many spaces to have high conservation values that allow their use as a resource to support other more traditional ones. Therefore, knowing the demographic and environmental changes of the municipalities bordering La Raya - A Raia and Bajo / Baixo Guadiana constitutes valuable information that leads to the territorial management of these peripheral areas.

2. Demographic evolution during the XXI century

The territorial scope established for this research includes a total of 38 municipalities, 15 of them in the Portuguese part (39.47%) and 23 (60.53%) in the Spanish (**Tables 1** and 2).

Portugal	Population 2019	Population 2011	Population 2001
Portalegre	22266	24789	25957
Elvas	20617	23000	23364
Vila Real de Sto. António	18819	19111	17981
Serpa	14339	15623	16705
Moura	13734	15091	16537
Reguengos de Monsaraz	10026	10815	11366
Vila Viçosa	7688	8306	8857
Mértola	6172	7226	8669
Campo Maior	7891	8449	8405
Castro Marim	6256	6695	6610
Alandroal	5028	5828	6554
Alcoutim	2202	2873	3743
Arronches	2840	3157	3381
Mourão	2453	2655	3204
Barrancos	1640	1822	1923
TOTAL inhabitants	141971	155440	163256

Source: National Institute of Statistics (Portugal).

Table 1.
Evolution of the number of inhabitants of the municipalities located in Portugal during the XXI century.

Spain	Population 2019	Population 2011	Population 2001
Badajoz	150702	151565	136319
Isla Cristina	21264	21903	18435
Ayamonte	20946	20763	17084
Olivenza	11963	12008	10739
Oliva de la Frontera	5137	5539	5933
Alburquerque	5343	5619	5605
Villanueva del Fresno	3397	3655	3487
Aroche	3073	3282	3446
Puebla de Guzmán	3073	3124	3220
Villanueva de los Castillejos	2820	2762	2672
La Codosera	2075	2320	2372
Alconchel	1680	1894	2069
Villablanca	2848	2916	2060
Rosal de la Frontera	1697	1913	1827
Encinasola	1305	1540	1772
Cheles	1176	1241	1323
Santa Bárbara de Casa	1035	1177	1312
Paymogo	1159	1308	1289
El Almendro	826	870	851
Valencia de Mombuey	744	809	811
San Silvestre de Guzmán	623	731	668
El Granado	516	567	653
Sanlúcar de Guadiana	409	440	381
TOTAL inhabitants	243811	247946	224328

Source: National Institute of Statistics (Spain).

Table 2.
Evolution of the number of inhabitants of border municipalities located in Spain during the XXI century.

If we carry out a detailed analysis of the municipalities located in the Portuguese territory, it stands out that from 2001 to 2019, 21,285 inhabitants have been lost (−13.03%). Especially striking are the cases of cities such as Portalegre, Elvas, or Moura with a population decline of between 2,000 and 3,000 inhabitants (**Table 1**). This shows a clear trend towards depopulation of these intermediate-scale municipalities.

However, if we analyze the 23 municipalities located in Spain, from 2001 to 2019 the population has grown by 19,483 inhabitants (+ 8.68%). However, if we analyze the data only from 2011, the last population census carried out in both countries concerning the 2019 data also shows a regressive trend (**Table 2**), which shows that as a whole the municipalities of The study area, especially as of 2011 on both sides of the borderline, shows a downward trend, although much more pronounced in the Portuguese part (**Tables 1** and **2**). These results coincide with the demographic predictions made by Mora, J. [34], where a trend towards the concentration of the population in large cities (Madrid and Lisbon, respectively) is confirmed, to the detriment of rural areas and even small cities and intermediate.

3. Natural heritage

The study area is part of the Southwest of the Iberian Peninsula, a territory characterized by a series of environmental elements and processes that are described below.

The primeval Mediterranean forest, with large tree formations of holm oaks, cork oaks, gall oaks and oaks, is considered the starting stage from which the set of plant formations presently present in these border areas originated. It is not known exactly what the appearance of this immense quercine forest looked like, but it would have a different appearance than the current one, much more uniform and continuous, covering practically the entire surface considered and interrupting only in the few and temporary clearings caused by the natural fires [35].

Human pressure began to act on this pristine nature, intensifying with population growth, advances in the use of tools, as well as the use of fire to obtain pastures and farmland. These labors favored the development of the cork oak for its resistance to fire and the holm oak and wild olive for its fruits.

Subsequently, the anthropic pressure increased, appearing uses such as those aimed at obtaining fuels (charcoal and firewood), beekeeping, or livestock, which took advantage of the resources from the bushes generated with the thinning of the forests.

The areas of little slope and more fertile, although with still mediocre soils for cultivation, were transformed into pastures, mainly since the Middle Ages, completely eliminating the scrub and reducing the density of the trees. The pasture area was increased through logging and fire, maintaining it with the help of livestock pressure, resulting in a sabanoid-looking system in which the trees had much greater bearing and fruit production than those of the previous stage. The reduction practiced on the woody biomass with the elimination of the scrub and the thinning and pruning of the trees increases the livestock productivity of the system, channeling the photosynthetic activity towards the pastures and acorns, easy to collect and store, and with the added value of possess a high energy power. The holm oaks are pruned with a low cross, around two or three meters, and with two or three main branches tending horizontally, thus favoring the production of grasses and fruits. In the case of cork oaks, and especially those located in Portugal, the trend is towards verticality, guiding the trees with long shafts and branches, achieving higher production and quality of the cork.

The current olive groves are a typical case of the evolution of the wild olive meadows as their fruit acquires greater importance than the acorn. These pastures disappeared as such through an initial grafting process that preserved a random distribution, being later replaced by crops with a regular arrangement that facilitated the work. At present, only some wild olive meadows have been located in Portuguese territory. If the tree stratum was completely removed and the fertility of the soil was adequate, crops were made possible, the typical rainfed ones prevailing due to the climatic characteristics of the southwest of the peninsula. In these cultivated fields, the original woody vegetation is maintained on the borders in the form of hedges and in small areas that do not allow agricultural practices, forming the typical checkered landscape of the cultivated areas.

In other cases, the high requirement by certain industries (paper, wood, food industry) of vegetable raw material of a different nature to the native, led, especially in some periods, to an intense repopulation policy that promoted profound landscape and socio-economic changes. This is the case of the implantation of non-native species that gave rise to the eucaliptus and pine forests.

As a consequence, the following environmental units currently predominate:

- Holm oaks: The primitive structure of holm oak forest and very dense understory has survived to this day with a greater or lesser degree of transformation. It has the appearance of a set of tall, medium-density trees accompanied by a more or less dense shrub layer, made up of a great variety of scrub species.

- Cork oaks: This forest develops mainly on deep siliceous soils with precipitation higher than those registered in the holm oak areas, generally distributed in an altitudinal band higher than that occupied by the latter and / or in shady areas, with a lesser degree of insolation, replacing the holm oak forest.

- Forest repopulation: The forestry policies carried out in the area in past decades, of greater intensity in Portugal than in Spain, sometimes favored the replacement of natural vegetation by other.

- Riparian forest: The riparian vegetation constitutes one of the most interesting systems, both from the landscape and floristic point of view. As a whole, riparian forests can constitute edaphic climaxes, or serial stages in the course of the ecological succession, in which the degree of proximity to the water course and its characteristics influence in a special way, that is, if they are continuous or intermittent currents.

- Shrub formations: Within them, the jarales (areas dominated by *Cistus* spp.) stand out, which with their dense cover does not allow the development of herbaceous vegetation in lower strata. Heaths (*Erica arborea* formations) also appear, characteristically on acid substrates and which develop especially after the disappearance of the cork and oak forests. Retamares (*Retama sphaerocarpa* areas) are linked and favored by livestock farms, presenting a lower density than jarales and heaths. Finally, we have the so-called nanomatorral, located in soils with high degradation and stoniness.

- Herbaceous formations: All the vegetation formations described, both arboreal and shrub, are always accompanied by a more or less relevant herbaceous layer. The pastoral interest of this stratum has caused the disappearance of the forest and the elimination of the scrub in large areas that do not have a clear agricultural vocation.

Wild fauna, linked to the plant matrix, constitutes a very relevant element both in the European and global context. The great variety of plant systems, added to the geomorphological and climatic diversity of the studied area, contribute to the fauna a whole range of biotopes and habitats of great value, which constitute enclaves for the survival of numerous animal species. This explains the numerous protected areas existing in this field of study originating both from the Natura 2000 Network and from the environmental planning of each country and which become important environmental resources.

4. Land use changes

The European Environment Agency (EEA) offers the CORINE Land Cover project (Coordination of Information Environment-CLC). This geographical database supplies land-uses in the European Union through polygon graphic features at a scale of 1:100,000 and with a minimum cartographic unit (MCU) of 25 hectares through polygonal graphic features. It also offers three hierarchical levels of information, consisting of the highest level of detail for 44 types of land-use classes in 1990, 2000, 2006, 2012 and 2018 [36]. The second layer of information used is also composed of polygonal graphic features containing the administrative delimitation of the 38 municipalities within the study area.

In this regard, the land-use information layer in 2000 and 2018, together with the administrative delimitation layer, were managed through a Geographic Information System (GIS) by using ArcGIS 10.5 software. Initially, the layer referring to the municipalities was used as a clipping mask on land-uses in 2000. Subsequently, the same procedure was repeated, but for the year 2018. In both cases, the resulting layer was a layer of polygons which contained for the year 2000 and 2018, the land-uses within each of the municipalities analyzed. In fact, within each polygon that delimited each municipality were the polygons related to land-uses according to CLC codification.

Then a field was then generated on each of the resulting layers. Then, in this field, the area of the polygons corresponding to the land-uses in hectares was geometrically calculated. Subsequently, through Structured Query Language (SQL), land-use records were selected for each municipality corresponding to 1) *Artificial surfaces*, 2) *Agricultural areas*, 3) *Forests and semi-natural areas*, 4) *Wetlands* and 5) *Water bodies* (CLC Level 1 nomenclature).

In the case of the layer corresponding to the year 2000, a total of 5,449 polygons were measured and by 2018 6,555 polygons were measured. Thus a greater diversification of land-uses was confirmed over the years.

In terms of land-uses in both 2000 and 2018, *Agricultural areas* predominated, increasing slightly from 64–68%. Therefore, more than half of the land is agricultural and this trend is maintained. The second main land-use corresponds to *Forests and semi-natural areas*, decreasing from 34% in 2000 to 28% in 2018. The remaining land-uses are quite a minority. Nonetheless, the increase of Water bodies is tripled, due to the creation of the Alqueva reservoir.

Regarding the land-uses analyzed by municipalities, the *Artificial surfaces* class only descends in 6 municipalities 4 in Portuguese territory and two Spaniards.

As for land-uses, taking into account the municipalities analyzed, the *Artificial surfaces* class only descends in 6 municipalities: 4 in Portuguese territory Alentejo, Alandroal, Campo Maior, Mértola, and Barrancos, and two of them are Spanish Cheles and Santa Bárbara de Casa. In all of them, there is a slight decrease. However, in Cheles this decrease is very pronounced going from 68.47 has to 44.49 has. There are also 4 municipalities, 2 of them Portuguese Portalegre and Alcoutim, and two Spaniards, Sanlúcar de Guadiana and Villanueva de los Castillejos, which do not vary. The rest of the 28 municipalities show an increase in the use of *Artificial surfaces*, highlighting those that experience an increase above 75% corresponding to the municipalities of Olivenza and Ayamonte.

As for *Agricultural areas*, the area goes down in 11 municipalities, highlighting the Spanish municipality of Isla Cristina whose decline is close to 25% and the Portuguese municipality of Reguengos de Monsaraz whose descent is close to 36%. The rest of the 27 municipalities increase in area for agricultural use, highlighting those that increase above 50% such as the Spaniards of La Codosera and Albuquerque, and the Portuguese municipality of Barrancos.

About *Forest and semi-natural* areas, there is a reduction in the area in 32 municipalities, 24 of which with a descent of more than 25%, highlighting Cheles that goes from 583.71 to 146.47 has and Alconchel that changes from 9,247.75 to 2,700.31 has. On the other hand, there are 6 municipalities with an increase in the area, which in Vila Real de Santo António reaches 25%, while in Campo Maior it reaches 66%.

Municipalities	1. Artif.	2. Agr	3. For.	4. Wet.	5. Wat.
Alandroal	337.3	35760.4	17887.0	0.0	703.8
Alburquerque	107.5	26093.0	44812.0	0.0	1283.5
Alconchel	40.2	20062.9	9247.8	0.0	206.3
Alcoutim	75.1	20043.3	38603.6	0.0	170.0
Almendro (El)	5.0	1700.7	12577.8	0.0	303.5
Aroche	26.0	12423.5	37443.8	0.0	0.0
Arronches	69.5	27003.5	3507.8	0.0	671.2
Ayamonte	246.1	2431.1	8604.2	2212.6	437.7
Badajoz	4375.1	128385.9	16657.5	0.0	790.6
Barrancos	49.7	8369.9	9566.0	0.0	0.0
Campo Maior	184.7	25197.7	196.3	0.0	60.9
Castro Marim	256.9	17221.8	9435.4	1272.0	953.6
Cheles	68.5	3969.1	583.7	0.0	182.3
Codosera (La)	63.8	1505.6	5372.7	0.0	0.0
Elvas	734.1	49075.4	5397.8	0.0	410.6
Encinasola	32.6	6915.5	10893.2	0.0	0.0
Granado (El)	0.0	1514.6	8022.1	0.0	157.7
Isla Cristina	295.2	2649.6	1050.1	690.0	108.0
Mértola	575.8	82931.6	43954.4	0.0	1028.3
Mourão	139.3	22621.9	5180.0	0.0	388.7
Moura	495.2	71961.7	23946.5	0.0	301.4
Oliva de la Frontera	75.9	9342.2	5538.8	0.0	0.0
Olivenza	126.7	33738.2	8737.2	0.0	493.3
Paymogo	26.7	10261.4	11082.7	0.0	0.0
Portalegre	0.0	8272.8	10083.8	0.0	0.0
Puebla de Guzmán	171.8	12247.3	19318.6	0.0	473.9
Reguengos de Monsar.	319.8	35648.4	9266.8	0.0	229.1
Rosal de la Frontera	46.3	6885.5	14090.7	0.0	0.0
San Silvestre de Guzm.	0.0	1100.2	3067.9	0.0	0.5
Sanlúcar de Guadiana	0.0	4001.8	4700.1	0.0	52.3
Santa Bárbara de Casa	37.6	5270.2	9422.1	0.0	0.0
Serpa	586.2	88163.4	19534.8	0.0	800.7
Valencia del Mombuey	0.0	3670.2	3834.4	0.0	0.0
Vila Real de Santo Ant.	248.7	1650.8	630.5	302.0	161.6
Vila Viçosa	987.8	14048.9	5131.2	0.0	0.0
Villablanca	36.6	2520.3	5278.1	0.0	30.2
Villanueva de los Cast.	0.0	605.4	371.1	0.0	0.0
Villanueva del Fresno	52.2	28746.4	7220.9	0.0	111.0
TOTAL	10893.5	834012.0	450249.2	4476.5	10510.7

Table 3.
Land-uses in 2000 measured in hectares according to the CLC level 1 nomenclature.

Municipalities	1. Artif.	2. Agr.	3. For.	4. Wet.	5. Wat.
Alandroal	292.8	41549.1	10734.6	0.0	2112.0
Alburquerque	216.5	43652.0	27106.6	0.0	1320.9
Alconchel	102.4	26112.1	2700.3	0.0	642.4
Alcoutim	75.1	17678.4	40940.8	0.0	197.8
Almendro (El)	11.8	3764.9	10082.3	0.0	728.0
Aroche	71.6	16654.9	33166.8	0.0	0.0
Arronches	86.0	27389.7	2856.8	0.0	919.5
Ayamonte	1025.3	3923.9	6071.4	2441.5	469.5
Badajoz	6297.7	135190.3	7949.4	103.1	668.8
Barrancos	44.5	10561.4	7379.7	0.0	0.0
Campo Maior	165.1	24633.0	590.5	0.0	251.1
Castro Marim	432.5	14464.9	12066.1	1221.2	955.1
Cheles	44.5	3919.5	146.5	0.0	693.2
Codosera (La)	80.8	2370.1	4491.2	0.0	0.0
Elvas	918.6	50545.9	3473.1	0.0	680.2
Encinasola	47.6	9144.4	8649.3	0.0	0.0
Granado (El)	38.6	3078.5	6406.9	0.0	170.4
Isla Cristina	400.5	3084.2	496.9	723.0	88.3
Mértola	569.5	77492.8	49404.3	0.0	1023.6
Mourão	146.0	20942.1	1989.4	0.0	5252.4
Moura	784.1	69087.3	22197.6	0.0	4635.7
Oliva de la Frontera	114.3	10561.3	4281.3	0.0	0.0
Olivenza	530.7	37336.3	4584.6	0.0	643.9
Paymogo	36.9	12493.0	8780.4	0.0	60.5
Portalegre	0.0	9052.3	9304.2	0.0	0.0
Puebla de Guzmán	247.5	17456.0	11248.7	0.0	3259.4
Reguengos de Monsar.	521.5	33717.4	6898.4	0.0	4326.7
Rosal de la Frontera	51.8	5545.5	15425.1	0.0	0.0
San Silvestre de Guzm.	25.1	2519.7	1623.0	0.0	0.8
Sanlúcar de Guadiana	0.0	5241.6	3468.0	0.0	44.7
Santa Bárbara de Casa	34.8	6377.1	8291.4	0.0	26.5
Serpa	703.7	87197.3	19351.7	0.0	1832.4
Valencia del Mombuey	28.4	5115.6	2360.5	0.0	0.0
Vila Real de Santo Ant.	435.8	1215.2	887.6	293.4	161.6
Vila Viçosa	1029.8	15636.6	3501.5	0.0	0.0
Villablanca	61.1	4328.3	3429.5	0.0	46.3
Villanueva de los Cast.	0.0	822.6	153.9	0.0	0.0
Villanueva del Fresno	105.2	32428.9	3039.1	0.0	557.3
TOTAL	15778.0	892283.9	365529.3	4782.0	31768.7

Table 4.
Land-uses in 2018 measured in hectares according to the CLC level 1 nomenclature.

Concerning *Wetlands*, it falls very slightly in two municipalities below 5% in Castro Marím and Vila Real de Santo António. In addition, there are also 33 municipalities where the *Wetlands* are not quantified. This type of coverage increases in 3 municipalities, and in two of them, Isla Cristina and Ayamonte, it is less than 10%. The most peculiar case is Badajoz with an increase of 100% from 0 to 103.03 has.

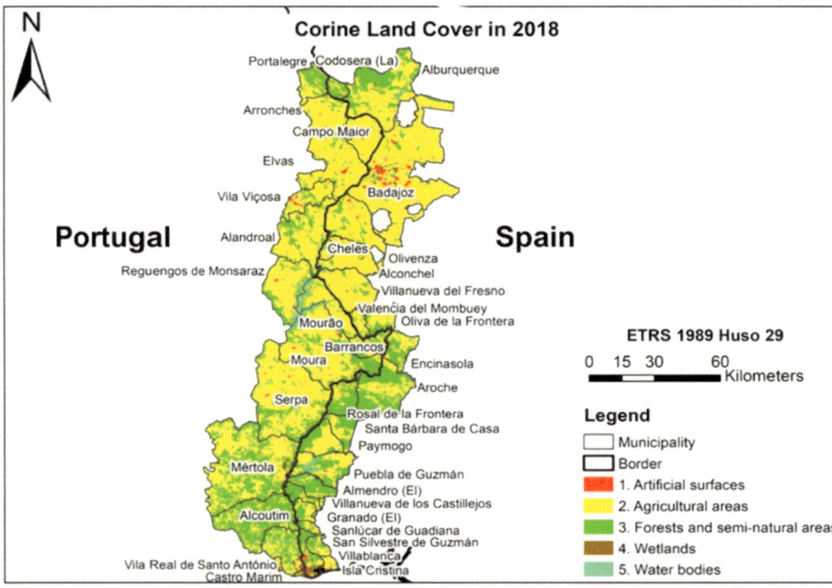

Figure 1.
CLC units in terms of land-use in 2018.

Municipalities	1. Artif.	2. Agr.	3. For.	4. Wet.	5. Wat.
Alandroal	−44.5	5788.7	−7152.4	0.0	1408.2
Alcoutim	0.0	−2365.0	2337.2	0.0	27.7
Arronches	16.5	386.2	−651.0	0.0	248.3
Barrancos	−5.2	2191.5	−2186.3	0.0	0.0
Campo Maior	−19.6	−564.7	394.2	0.0	190.2
Castro Marim	175.6	−2756.9	2630.7	−50.9	1.4
Elvas	184.6	1470.5	−1924.7	0.0	269.6
Mértola	−6.3	−5438.8	5449.9	0.0	−4.7
Mourão	6.7	−1679.8	−3190.6	0.0	4863.7
Moura	289.0	−2874.4	−1748.9	0.0	4334.3
Portalegre	0.0	779.5	−779.6	0.0	0.0
Reguengos de M.	201.7	−1931.0	−2368.4	0.0	4097.7
Serpa	117.5	−966.1	−183.1	0.0	1031.7
Vila Real de S. António	187.1	−435.6	257.1	−8.7	0.0
Vila Viçosa	42.0	1587.6	−1629.7	0.0	0.0
TOTAL	1145.2	−6808.2	−10745.5	−59.5	16468.1

Table 5.
Differences in Portugal in land-uses (hectares) between 2000 and 2018 according to the CLC level 1 nomenclature. Negative values indicate losses in 2018 compared to 2000.

Municipalities	1. Artif.	2. Agr.	3. For.	4. Wet.	5. Wat.
Alburquerque	109.0	17559.0	−17705.4	0.0	37.4
Alconchel	62.2	6049.2	−6547.4	0.0	436.1
Almendro (El)	6.8	2064.2	−2495.5	0.0	424.5
Aroche	45.6	4231.4	−4277.0	0.0	0.0
Ayamonte	779.2	1492.8	−2532.8	228.9	31.9
Badajoz	1922.6	6804.4	−8708.2	103.1	−121.9
Cheles	−24.0	−49.7	−437.2	0.0	510.9
Codosera (La)	17.0	864.5	−881.5	0.0	0.0
Encinasola	15.0	2228.9	−2243.9	0.0	0.0
Granado (El)	38.6	1563.9	−1615.2	0.0	12.7
Isla Cristina	105.3	434.6	−553.2	33.0	−19.7
Oliva de la Frontera	38.4	1219.1	−1257.5	0.0	0.0
Olivenza	404.0	3598.1	−4152.7	0.0	150.5
Paymogo	10.2	2231.6	−2302.3	0.0	60.5
Puebla de Guzmán	75.7	5208.7	−8069.9	0.0	2785.5
Rosal de la Frontera	5.5	−1340.0	1334.5	0.0	0.0
San Silvestre de G.	25.1	1419.5	−1444.9	0.0	0.3
Sanlúcar de Guadiana	0.0	1239.8	−1232.2	0.0	−7.6
Santa Bárbara de Casa	−2.8	1106.9	−1130.6	0.0	26.5
Valencia del Mombuey	28.4	1445.5	−1473.9	0.0	0.0
Villablanca	24.5	1808.0	−1848.6	0.0	16.1
Villanueva de los C.	0.0	217.2	−217.2	0.0	0.0
Villanueva del Fresno	53.0	3682.5	−4181.8	0.0	446.3
TOTAL	3739.3	65080.1	−73974.4	365.0	4789.9

Table 6.
Differences in Spain in land-uses (hectares) between 2000 and 2018 according to the CLC level 1 nomenclature. Negative values indicate losses in 2018 compared to 2000.

Finally, the land-use classified as *Water bodies* is registered by only 4 municipalities where its area is reduced: Mértola, Sanlúcar de Guadiana, Badajoz and Isla Cristina. The latter accumulates a drop above 22%. In another 11 there is no area difference; in 10 of them due to the fact that there is no hectare corresponding to this land-use neither in 2000 nor 2018. The increase in area is recorded in 23 municipalities, 8 of which above 75%, highlighting Paymogo and Santa Bárbara de Casa with an increase of 100% (**Tables 3** and **4**).

Subsequently, in order to identify the situation of land-uses and patterns in the territory, thematic maps were generated for each of the years analyzed and taking into account the classification of land-uses for CLC level 1.

Figure 1 shows that the predominant land-use in 2018 corresponds to *Agricultural areas*, although in the border area to the south and in the north the predominant coverage corresponds to *Forest and semi-natural areas*, with minor spots scattered throughout the rest of the territory. As for Artificial surfaces, there seems to be a greater concentration in Badajoz, due to the fact that it is the most populated municipality within the area analyzed. As for the rest of the land-uses, due to being a minority in quantity, they are practically not observable.

Tables 5 and **6** shows the differences produced between 2000 and 2018. The loss of the total area of the *Forests and semi-natural areas* in both the territory of Spain and Portugal is the most prominent. The latter country also lost agricultural land, more than 6,800 ha. These losses are mainly compensated by the increase in the *Agricultural areas* and *Water bodies* classes on the Spanish side, while in the Portuguese side, what increases the most is the territory of *Water bodies* due to the appearance of the great reservoir of Alqueva, which closed its floodgates in 2002. When these variations are checked by municipalities in the Spanish territory, there are significant variations. In this way, there is a concentration of losses in forested areas and an increase in the other classes of the CLC. On the contrary in Portuguese municipalities, there is greater variability in losses and gains of areas in the five classes.

5. Heritage and border tourism

5.1 The Guadiana border as a tourist itinerary

Our territorial area of Bajo Guadiana is also a tourist space with a multitude of resources and various products. It contains a territory rich in historical-artistic and natural heritages and that could also benefit from border tourism. In this sense, our attention is going to focus on this "border tourism", a segment or tourism product that is recent in time, still poorly defined and that is very transversal to others (rural, ecotourism, agrotourism, gastronomic, etc.) that occur in border territories [37].

We are located in a border section with an interesting history that now allows a free movement of movements and travelers, apart from the recent problems caused by Covid 19. The border itself could become a destination and the main attraction for a very diverse tourist demand, eager to consume the history and heritage of the frontier territories.

Hernández [38] distinguishes between cross-border tourism and tourism on the border. In the first, the territory to be visited covers a more or less wide space, of transition and somewhat lax limits on both sides of the border and where the border fact has been felt. On the contrary, tourism on the border is characterized by the fact that the destination is the same border line and the tourist activity is focused on visiting neighboring populations that belong to different states. The most interesting examples are those that occur between neighboring towns, but from different countries, where the border is a river or a customs office. The most eloquent cases are those of Alcoutim and Sanlúcar de Guadiana or that of Badajoz and Elvas.

In the case of our field of study, the border has a linear sense. Somehow it is a route that runs parallel to the political dividing line between Spain and Portugal and that tourists can follow or cross looking for their border heritage and, at the same time, other resources: monuments, landscape, gastronomy, cultural events, sports practices, etc. The border, as a destination, thus not only becomes an attraction, but is also configured into a tourist itinerary. In a way, our border territory could act or convert in an analogous way to other well-known tourist-cultural itineraries or routes: historical roads, pilgrimage routes, routes of heritage cities, etc. This situation can develop in integrated borders, such as that of the Iberian line (La Raya), with the particularity of being an itinerary that for many centuries has been a political barrier (the oldest political border in Europe) and that only in the last decades it has been able to permeate to a greater degree for the transit of its inhabitants and travelers. At present it is an integrated border from the perspective of mobility with numerous new bridges, with abolished customs and with territorial cooperation policies (Interreg) that try to revalue the cultural and natural heritage and activate leisure and tourism activities.

So far, a group of researchers (where the Turfront project stands out) have worked to make this product or typology of "border tourism" official, motivated by the desire that it be assumed by the different tourism policies and programs at regional and local scales. This new concept would encompass as such the movements of travelers and tourists and the tourist products that are supported as a destination mainly on the border, its history and heritage.

In our territory of Bajo Guadiana, which also extends to the whole of the Iberian Line, there is a rich history that has generated an outstanding historical and cultural legacy, to which the environment is added, and that could now be the main attractive or revulsive resources for a major and varied tourist activity. And with the particularity also of being a somewhat unexplored and little-known territory for mass tourism.

However, the task of inserting or consolidating these basically rural spaces, such as itineraries and international tourist destinations, is not easy. Campesino [39] refers that most of the time the natural and cultural heritages of the border are still "potential resources with attractions and expectant profitability" that have not yet been transformed into border tourism products. Undoubtedly, the great challenge of this type of incipient destinations is to make the leap from being spaces of complementary offer to being true destinations with an offer of sufficient and profitable accommodation for the hiker or accidental traveler to become a true tourist of the border.

5.2 Guadiana cultural heritage and tourism

We conceived the Guadiana River and the border of la Raya as a tourist itinerary from the South Atlantic Coast to a more unknown interior but rich in resources and historical-artistic and natural heritage. Therefore, we will then give a brief description of these potentials and resources by river areas from south to north.

At the southern apex, as a great linear and supporting element appears an unknown international section of the Guadiana River, between meanders and banks of unexpected and surprising attractiveness. We believe that this river has remarkable potential for tourist use (river walks and sports activities), at least in its approximately 60 kilometers of international and navigable stretch, from the mouth to its encounter with its tributary of Chanza. The river ports and fortifications of Alcoutim (Algarve), Mértola (Baixo Alentejo) and Sanlúcar de Guadiana (Huelva province) and the small jetties of Pomarão, Foz de Odeleite and Guerreiros, on the Portuguese shore, and Puerto La Laja in the municipality of El Granado stand out. They are joined by the larger ports of Ayamonte, Isla Cristina and Vila Real de Santo Antonio, almost at the same mouth.

With proper name, border towns such as Ayamonte, Sanlúcar de Guadiana and Aroche stand out, on the Spanish side, and Castro Marim, Alcoutim, Mértola and Serpa, in Portuguese lands stand out too. All of them are historical ensembles protected by their outstanding architectural cultural heritage. Likewise, an interesting network of trails, greenways and viewpoints has been built for the traveler who wants to discover nature and history. In addition, a set of mining testimonies, museums and interpretation centres mark this territory of Lower Guadiana. This cultural heritage is joined by spaces of marshes (zapales in Spanish) and Mediterranean forest that have been protected (Guadiana Valley Natural Park).

In this short descriptive route, we distinguish a second section and that refers to the Alqueva reservoir and its adjacent territory. In addition to the agricultural and energy objectives and uses, this territory has become due to its enormous dimensions (almost an inland sea) a new and important tourist destination [40]. Various marinas and tourist villages have been built and several active leisure companies have appeared, with wide possibilities of river and land tours. Lake Alqueva has also been the world's first certified Starlight destination in 2011 under the name Dark Sky Alqueva Reserve.

These tourist uses around the water have brought the traveler and tourist closer to the cultural heritage, mainly of fortifications, of Portuguese towns such as Moura, Mourão, Reguengos de Monsaraz and Alandroal. Their castles and walls are impressive, they are witnesses of other times of wars and mistrust. On the Spanish side, the shores of the great lake reach Villanueva del Fresno, Cheles, Olivenza and Alconchel. Very outstanding is the historical-artistic complex in Olivenza. Further away from the shore of Alqueva there is a triangle formed by the towns of Barrancos, Encinasola, Valencia del Mombuey and Oliva de la Frontera, where its grasslands and other various defensive strongholds stand out.

A third zone, further north, would encompass the municipalities of Vila Visosa, Elvás, Campo Maior, Arronches and Portalegre; and for Spain, the municipalities of Badajoz, La Codosera and Alburquerque. Here are the most populated and urban centres (Badajoz and Portalegre). It also highlights its cultural heritage, where Elvás stands out, whose walled fortress has been declared a World Heritage Site since 2012. On the Portuguese side, the Sierra de San Mamede Natural Park highlights which reaches the border, facing lands in the province of Caceres. As for Spain, the historical and heritage legacy of Badajoz and Alburquerque, with walled enclosures, is equally outstanding.

The existence of the Eurocity Badajoz-Elvas-Campo Maior (Eurobec), with about 190,000 inhabitants, is noteworthy. Besides, since its formal creation in 2018, this Eurocity aims to strengthen border tourism and combine various services and facilities [41]. The Baluartes project has been launchedas part of the revaluation of the fortifications of the two cities. Likewise, work is also being done for the future integration of the municipality of Olivenza.

6. Conclusions: final reflections and proposals

The territory that hosts the border municipalities of the first line of the Guadiana basin is characterized by being deeply rural (not to be confused with agrarian), both due to the size of its localities, only one exceeds 150,000 inhabitants (Badajoz) and nine exceed 10,000 residents, which is the threshold that the INE considers "city": 6 in Portugal and 3 in Spain, although they host 78.97% of the total.

There is evidence of a gradual decline throughout the millennium of the Portuguese municipalities, except Vila Real de Santo António, due to its tourist beach function. In the Spanish part, growth is observed in most of the nuclei between 2001 and 2011, however, they all enter a regressive phase from 2011 to the present, except for Badajoz and Ayamonte.

We are witnessing a demographic decline in 94.4% of the municipalities and 92.7% of the border area, most of the localities being condemned to an irreversible scenario if the enormous natural and cultural potentials are not taken advantage of.

Regarding land uses, two uses are predominant. In the first place, the one corresponding to agricultural use, since it occupies more than half of the analyzed area. It even increased between 2000 and 2018, consolidating its hegemony. Second, the land use referred to Forests and semi-natural areas predominates, occupying approximately one third. However, the trend is contrary to the previous land use, since the area occupied by this type of land decreases slightly, thus progressively losing prominence over the years.

The rest of the land uses are a minority, although the notable increase in the Waterbodies class stands out. As a consequence of the creation of the Alqueva reservoir in 2002. This effect is more noticeable in the municipalities closest to the reservoir, Alandroal, Mourão, Moura, and Reguengos de Monsaraz in Portugal, and Alconchel, Cheles, Olivenza, and Villanueva del Fresno in Spain. Among all of them, Mourão stands out, since the use of the land water bodies now occupies 17% of the total area of the municipality and Cheles, where it now has 10%.

The evolution of the population of this territory is worrying, characterized by a general decline. Although in some municipalities there was a small rebound towards the middle of the period studied, almost a decade later they lose inhabitants again. Only the two coastal municipalities in Spain increase demographically, a situation that is not reflected in the Portuguese coastal municipalities. Consequently, the effect of the construction of the Alqueva reservoir has not been felt as a catalyst that contributes to the fixation of the population. Moreover, in the Portuguese area, the loss has been more pronounced. Precisely in the Baixo Alentejo the impact should have been greater since almost all of the newly irrigated areas have been built there. Undoubtedly, the demographic decline that was already dragging this territory outweighs the new economic exceptions of the aforementioned reservoir and where irrigated agriculture stands out.

We find a space with different geoenvironmental units whose common denominator is the Guadiana River and the border factor. We are facing a border culture and history linked to wars, the continuous construction, and deconstruction of the border, the unique relationships of attraction and repulsion with the neighbor, smuggling, mining, and a singular ethnography or immaterial culture. It is these endogenous resources (mining, river, forest, agricultural, hunting, landscape, historical, heritage ...) that should guide any policy or plan aimed at generating development processes. And ideally, these planning and intervention policies are comprehensive and cross-border.

A territory, that of Bajo Guadiana, with very high landscape and heritage values that show potentialities that are still very little exploited by tourism, far from being even a destination like other well-known cultural itineraries. However, the sum of the patrimonial factors, the better accessibility, and the cooperation policies have already allowed the emergence of some tourist offers and services in these border towns when, until recently, nothing existed.

The greater connection between the two countries and the permeabilization of the border, once an almost insurmountable barrier, has undoubtedly substantially changed the flows throughout the southwest of the peninsula. Therefore, at the level of communication infrastructures, now there are conditions to create or promote a linear tourist itinerary and, at the same time, cross border tourism that allows the traveler and tourist to travel and cross this part of the Iberian border enjoying its resources and attractive.

In a very concise way, we believe it is appropriate to point out some proposals for the promotion of border tourism and this border area:

- Creation of a brand for border tourism in this territory. It could contain the names of "Guadiana" and "Frontera".

- Enhancement of cross-border tourism products. In this sense, and as an example, products/routes such as routes of castles and fortifications or the offer of active leisure around the Guadiana river could be reinforced.

- Change of mentality towards a more touristy local culture in which quality is committed and, within the different tourist products, the border type.

We think that the Guadiana and the Border can become a necessary tourist itinerary, with offers geared towards tourist demands that seek new experiences in little-known destinations and around tourist typologies linked to rural areas, heritage, and history. A greater revaluation of the cultural and natural heritage and an increase in tourism could be a complementary solution to curb the phenomenon of rural depopulation that plagues this territory and most of the Portuguese-Spanish line. It would be a new product, border tourism, perfectly compatible and transversal to others in these border areas.

It will be necessary to see the behavior of the population in the next decade to draw more conclusions from the variations it suffers and from the activities and policies that are carried out, analyzing their value for both economic and environmental sustainability.

Acknowledgements

The authors wish to acknowledge funding for this research work from the VI Regional Research Plan and the Regional Government of Extremadura and the European Regional Development Fund (ERDF), associated with financing the research group Sustainable Development and Territorial Planning (GR18052), the Environmental Resources Analysis Research Group (GR18054), Functional Study of Mediterranean Ecosystems (GR18078) and Institute of Local Development (HUM260) from the University of Huelva.

Author details

José Manuel Naranjo Gómez[1,2,3], Jacinto Garrido Velarde[1,3,4], José Martín Gallardo[5], José Manuel Jurado Almonte[6,7], Julian Mora Aliseda[1,3] and José Cabezas Fernández[1,2,8]*

1 INTERRA - University Research Institute for Sustainable Territorial Development, University of Extremadura, Spain

2 VALORIZA - Research Centre for Endogenous Resource Valorization, Polytechnic Institute of Portalegre (IPP), Portugal

3 DESOSTE - Sustainable Development of the Territory Research Group, University of Extremadura, Spain

4 Department of Social Sciences, Languages and Literatures, University of Extremadura, Spain

5 Functional Studies of Mediterranean Ecosystems, University of Extremadura, Spain

6 Department of History, Geography and Anthropology, University of Huelva, Spain

7 Institute of Local Development, University of Huelva, Spain

8 ARAM - Environmental Resources Analysis Research Group, University of Extremadura, Badajoz, Spain

*Address all correspondence to: jocafer@unex.es

References

[1] Konrad V, Nicol HT. Border Culture, the Boundary Between Canada and the United States of America, and the Advancement of Borderlands Theory. Geopolitics. 2011;16(1):70-90. DOI: 10.1080/14650045.2010.493773

[2] Laine JP. A historical view on the study of border. In: Sevastianov SV, Laine JP, Kireev AA, editors. Introduction to Border Studies. Vladivostok: Far Eastern Federal University; 2015. p. 14-32.

[3] Allen W. "I am From Busia!": Everyday Trading and Health Service Provision at the Kenya–Uganda Border as Place-Making Activities. Journal of Borderlands Studies. 2013; 28(3):291-306. DOI: 10.1080/08865655.2013.862756

[4] Konrad, V. Toward a Theory of Borders in Motion. Journal of Borderlands Studies. 2015;30(1):1-17. DOI: 10.1080/08865655.2015.1008387

[5] Batista T, Rodríguez FC, Carriço C, Pinto-Gomes C, Fernández-Pozo L, Gómez JM, Martín Gallardo J, Loures L, Cabezas J. (2020). Benefits of a Spatial Data Infrastructure on the Sustainability of a Southwestern European Territory. In: Castanho RA, editors. Cross-Border Cooperation (CBC) Strategies for Sustainable Development. Hershey, USA: IGI Global; 2015. p. 1-29. DOI: 10.4018/978-1-7998-2513-5

[6] Brunet-Jailly E. Special Section: Borders, Borderlands and Theory: An Introduction. Geopolitics. 2011;16:1-6. DOI: 10.1080/14650045.2010.493765

[7] Foucher M. Fronts et frontiers. Un tour de monde geopolitique. París, FR: Faeard; 1988. 691 p.

[8] Barros M, González A. Língua e Cultura na Fronteira Norte-Sul. Bibliografia. Mértola PT: Campo Arqueológico de Mértola; 2015. 389 p. Available from: http://www.camertola.pt [Accessed: 2020-10-08]

[9] Sideway JD. Rebuilding bridges: A critical geopolitics of Iberian transfrontier cooperation in a European context. Environment and Planning D: Society and Space. 2001;19(6):743-778. DOI: 10.1068/d223t

[10] Van Houtum H, Struver A. Borders, strangers, doors and bridges. Spaces & Polity. 2002;6(2):141-146. DOI: 10.1080/1356257022000003590

[11] Lofgren O. Regionauts: The transformation of cross-border regions in Scandinavia. European Urban and Regional Studies, 2008;15(3):195-209. DOI: 10.1177/0969776408090418

[12] Mora J, Garrido J. Demografía y medio ambiente de la cuenca hispano-lusa del río Tajo. Beau Bassin, Mauritius: OmniScriptum Publishing Group; 2017. 100 p.

[13] Levers C, Müller D, Erb K, Haberl H, Rudbeck Jepsen M, Metzger MJ, Meyfroidt P, Plieninger T, Plutzar C, Stürck J, Verburg PH, Verkerk PJ & Kuemmerle T. Archetypical patterns and trajectories of land systems in Europe. Regional Environmental Change. 2016;18:715-732. DOI: 10.1007/s10113-015-0907-x

[14] Munroe DK, van Berkel DB, Verburg PH & Olson JL. Alternative trajectories of land abandonment: causes, consequences and research challenges. Current Opinion in Environmental Sustainability, 2013;5:471-476. DOI: 10.1016/j.cosust.2013.06.010

[15] Aide TM, Clark ML, Grau HR, López-Carr D, Levy MA, Redo D, Bonilla-Moheno M, Riner G, Andrade-Núñez MJ, Muñiz M.

Deforestation and reforestation of Latin America and the Caribbean (2001-2010). Biotropica, 45:262-271. DOI: 10.1111/j.1744-7429.2012.00908.x

[16] Mantero G, Morresi D, Marzano R, Motta R, Mladenoff DJ, Garbarino M. The influence of land abandonment on forest disturbance regimes: a global review. Landscape Ecology. 2020;35:2723-2744. DOI: 10.1007/s10980-020-01147-w

[17] Moreira F, Rego FC, Ferreira PG. Temporal (1958-1995) pattern of change in a cultural landscape of northwestern Portugal: implications for fire occurrence. Landscape Ecology. 2001;16:557-567. DOI: 10.1023/A:1013130528470

[18] Farina A. Principles and Methods in Landscape Ecology: Toward a Science of Landscape. Dordrecht, The Netherlands: Springer; 2006. 412 p. DOI: 10.1007/978-1-4020-5535-5

[19] Cleland D., Crow TR, Saunders SC, Dickmann DI, Maclean AL, Jordan JK, Watson RL, Sloan AM, Brosofske KD. Characterizing historical and modern fire regimes in Michigan (USA): A landscape ecosystem approach. Landscape Ecology. 2004;19:311-325. DOI: 10.1023/B:LAND.0000030437.292 58.3c

[20] González-Bernáldez F. Ecological consequences of the abandonment of traditional land use systems in central Spain. Options Mediterranéennes. 1991;15:23-29.

[21] von Droste B, Plachter H, Rössler M. (eds). Cultural Landscapes of Universal Value — Components of a Global Strategy. Jena, Germany: Gustav Fischer. in association with UNESCO, Paris; 1995. 464 p.

[22] Olsson EGA, Austrheim G, Grenne SN. Landscape changes patterns in mounains, land use and environmental diversity, Mid-Norway 1960-1993. Landscape Ecology. 2000;15:155-170. DOI: 10.1023/A:1008173628016

[23] Beilin R, Lindborg R, Stenseke M, Pereira HM, Llausàs A, Slätmo E, Cerqueira Y, Navarro LM, Rodrigues P, Reichelt N. Analysing how drivers of agricultural land abandonment affect biodiversity and cultural landscapes using case studies from Scandinavia, Iberia and Oceania. Land Use Policy. 2014;36:60-72. DOI: 10.1016/j.landusepol.2013.07.003

[24] Pulido Fernández M, Lavado Contador JF, Schnabel S, Gómez Gutiérrez Á, Lozano-Parra J. Changes in land management of Iberian rangelands and grasslands in the last 60 years and their effect on vegetation. In: Sebata A, editor. Vegetation. London, UK: IntechOpen; 2017. p. 24-42. DOI: 10.5772/intechopen.72490

[25] COPERNICUS. CORINE Land Cover. 2020. Available from: https://land.copernicus.eu/pan-european/corine-land-cover [Accessed: 2020-09-11]

[26] Feranec J, Jaffrain G, Soukup T, Hazeu G. Determining changes and flows in European landscapes 1990-2000 using CORINE land cover data. Applied Geography. 2010;30:19-35. DOI: 10.1016/j.apgeog.2009.07.003

[27] Castanho RA, Lousada S, Naranjo JM, Escórcio P, Cabezas J, Fernández-Pozo L, Loures L. Dynamics of the Land Use Changes and the Associated Barriers and Opportunities for Sustainable Development on Peripheral and Insular Territories: The Madeira Island (Portugal). In: Loures LC, editor. Land Use - Assessing the Past, Envisioning the Future. London, UK: IntechOpen; 2018. DOI: 10.5772/intechopen.80827.

[28] Diaz-Pacheco J, Gutiérrez J. Exploring the Limitations of CORINE

Land Cover for Monitoring Urban Land-Use Dynamics in Metropolitan Areas. Journal of Land Use Science. 2014;9(3):243-259. DOI: 10.1080/1747423X.2012.761736

[29] Corbau C, Zambello E, Rodella I, Utizi K, Nardin W, Simeoni U. Quantifying the impacts of the human activities on the evolution of Po delta territory during the last 120 years. Journal of Environmental Management. 2019;232:702-712. DOI: 10.1016/j.jenvman.2018.11.096

[30] Feranec J, Hazeu G, Christensen S, Jaffrain G. Corine land cover change detection in Europe (case studies of the Netherlands and Slovakia). Land Use Policy. 2007;24(1):234-247. DOI: 10.1016/j.landusepol.2006.02.002

[31] Kucsicsa G, Popovici EA, Bălteanu D, Grigorescu I, Dumitraşcu M. Future land use/cover changes in Romania: regional simulations based on CLUE-S model and CORINE land cover database. Landscape and Ecological Engineering. 2019;15:75-90. DOI: 10.1007/s11355-018-0362-1

[32] Patterson J, Schulz K, Vervoort J, van der Hel S, Widergerg O, Adler C, Hurlbert M, Anderton K, Sethi M. Exploring the governance and politics of transformations towards sustainability. Environmental Innovation and Societal Transitions. 2017;24:1-16. DOI: 10.1016/j.eist.2016.09.001

[33] Tàbara JD, Cots F, Pedde S, Hölscher K, Kok K, Lovanova A, Capela Lourenço T, Frantzeskaki N, Etherington J. Exploring Institutional Transformations to Address High-End Climate Change in Iberia. Sustainability. 2018;10(1):161. DOI: 10.3390/su10010161

[34] Mora Aliseda J, Mora C, Garrido Velarde J. Demographic structures and the geographical distribution of the municipalities of the Guadiana hydrographic basin: Iberian Peninsula. WSEAS Transactions on Environment and Development. 2018;14:170-181.

[35] González-Bernáldez F. Ecología y Paisaje. Madrid: H. Blume; 1981 256 p.

[36] CORINE (CLC). 2020. Available from: https://www.eea.europa.eu/publications/COR0-landcover [Accessed: 2020-10-05]

[37] Jurado, J.M. y Pazo, F.J. (2018). "La Frontera como Recurso Turístico. Posibilidades del Bajo/Baixo Guadiana". Revista de Estudios Andaluces, vol. 35 (1), Universidad de Sevilla, 1-34.

[38] Hernández-Ramírez J. Turismo en la frontera: patrimonialización y cooperación transfronteriza en una periferia de la Unión Europea. Etnográfica. 2017;21(2):385-409. DOI: 10.4000/etnografica.4940

[39] Campesino AJ, Jurado JM, editors. Turismo de Frontera (III). Productos turísticos de la Raya Ibérica. Huelva, SP: Servicio de publicaciones, Universidad de Huelva; 2014. 351 p.

[40] Martins J, Jurado JM, Jaraíz J. Planeamento e oportunidades para o turismo no território de Alqueva. In: Mora J, Garrido J, Castanho RA, editors. Planeamiento sectorial: recursos hídricos, espacio rural y fronteras. Cizur Menor, SP: Aranzadi (Thompson Reuters); 2018. p. 235-250.

[41] Jurado JM, Pazo FJ, Castanho A. Eurocities of the Iberian Borderland: A Second Generation of Border Cooperation Structures. An Analysis of Their Development Strate-gies. Sustainability. 2020;12(16)6438. DOI: 10.3390/su12166438

Chapter 3

Common Regional Development Strategies on Iberian Territories - A Framework for Comprehensive Border Corridors Governance: Establishing Integrated Territorial Development

Ana Vulevic, Rui Alexandre Castanho, José Manuel Naranjo Gómez, José Cabezas, Luis Fernández-Pozo, Jacinto Garrido Velarde, José Martín Gallardo, Sérgio Lousada and Luís Loures

Abstract

Although the Trans-European Transport Network (TEN-T) has made a considerable contribution to increasing transport infrastructure and improving transport services, it still often seems that border regions have reduced accessibility compared to central regions. The Iberian border regions of Alto-Alentejo (Portugal) and Badajoz (Spain) were analyzed. Studies conducted in the Iberian territories confirm some of these regions' unsustainability and the non-impact on the inhabitants' quality of life. Also, a few questions arise: *Can we verify the change in methods and measures used by development strategies? Is it necessary to acquire a more comprehensive view of corridor flows and interactions through common regional development strategies for managing development corridors? Will we influence the change of the methodological approach to be strategic and that it does not imply a corridor only in terms of the flow and mobility of people and goods, but also the search for the causes of movements in growth corridors?* So, infrastructure and connectivity planning, regional economic development, and land use planning are critical contexts in developing territorial policies in the planning of development of corridors and their integration into development processes. This study proposes a framework for strategic corridor governance to increase accessibility and regional equity.

Keywords: regional development, corridor, accessibility, governance, sustainable planning

1. Introduction

Nowadays, Cross-Border Cooperation (CBC) strategies and territorial governance arrangements are essential to use territories to use external funding opportunities to finance their vision of their community or territory's future. However, it is not so simple to apply in practice. Although the Trans-European Transport Network (TEN-T) has made a considerable contribution to increasing transport infrastructure and improving transport services, it still often seems that border regions have reduced accessibility compared to central regions. The Iberian border area along the border regions of Alto Alentejo (Portugal) and Badajoz (Spain) was analyzed, where border regions still appear to be less accessible compared to the central regions despite the considerable contributions of the Trans-European Transport Network (TEN-T) in terms of equipped with transport infrastructure.

In order to activate all the potentials of border areas in recent decades, EU policies are aimed at increasing cross-border cooperation [1–4].

Common regional development strategies on Iberian territory can be a new framework for comprehensive border corridors governance and situating integrated territorial development. The present research will be presented some hypotheses and new ways "*to generate policy-relevant insights into the interactions and flows occurring within growth corridors*" for a well-planned long-term sustainable territory.

In their discussions on cross-border planning, Peironi and Denert in [5] emphasize the importance of EU regions having their institutional policy towards border regions.

These areas are very often internal peripheries, and their activation can be facilitated by various cross-border interactions that can be planned within common strategies [1–3, 5–8].

Furthermore, building a strategy should identify the issues associated with an area's inner peripheralization and understand the triggers and drivers [9]. Here we made a summary of the data corridor and policy and recommendations measure that can be useful for corridor collaboration, territorial development, and comprehensive corridor governance.

2. Common regional development strategies and governance across borderlands

A policy to enhance or modify the process of peripheralization needs a centered intervention philosophy and relevant combined implementation structures formed within the local, territorial and national governance methods, and demands to be practical, using current policy frameworks and convenient "levers" to perform influence. Besides, a particular focus should then be set on those means thriven by ineffectual connectedness of some form because this differentiates an inner perifieries (IP) from other typologies of the bordering region. The difficulty then is to obtain drivers that can cause a development at an advanced stage and turn around the "spiraling-down" methods of interior boundaries. Although each of the three IP thoughts has distinguishing features, they have, in general, a focus on the requirement to improve various kinds of connectedness [9].

Diverse investigations have analyzed the different kinds of multi-level governance in terms of European integration [10–12]. Multi-level governance formats are necessary for vertical and horizontal coordination between decision-making bodies at different levels and sector policies to manage links between different governance processes, develop multisectoral coordination on local, regional, national, and European actors [6].

Territorial management systems are consequently essential to maintain the development of regional administrations' organizational and flexibility capabilities, both individually and collectively. So, in a period of enhanced mobilities, the above-mentioned organizations are required to produce a living space capable of handling the in-and outflows of persons - i.e., transitional economy; thus, it is possible to valorize human and financial capital of its citizens - i.e., residential economy; also, to mobilize extra-local networks to create new knowledge encouraging community growth. A relevant characteristic of territorial governance is the regions' capacity to manage external funding possibilities to finance their perception of their society or territory's future [13].

Territorial strategies set specific indicators to check if the regional development is on track. Firstly it should operate primarily through "soft" means, such as the provision of regional analysis and collaboration.

Almost all management levels influence the development of the corridor, so the planning of the growth corridor requires a comprehensive approach to data and their sources to better understand the mutual interactions. Consequently, targeted big data analysis and EGCs were designed to establish the evidence-based plan in European growth corridors joining the TEN-T transport infrastructure network [14].

For a broader perception of the synergies within the growth corridor, traditional data references are sufficient, yet more comprehensive, new sources for more comprehensive corridor management [14].

In this regard, the issue is seen from different perspectives for managing growth corridors, carrying into account three coinciding dimensions of development (**Figure 1**) [14].

Contextually, we should consider three dimensions: (a) the physical dimension indicating transportation progress and urbanization; (b) the digital dimension recognizing digital flows and interplays as a significant part of corridor functions; and (c) the social dimension surveying the corridor as a framework for social synergy [14].

So, it is fundamental to progress from data management to data management based on perspicacity received from data and designing the basis for regional development management.

In this regard, **Table 1** presents the categorization of critical flows and interactions, which are extremely important to corridor governance and decision-making, with models of the kinds of data sets that appeal to each type of flow and interaction.

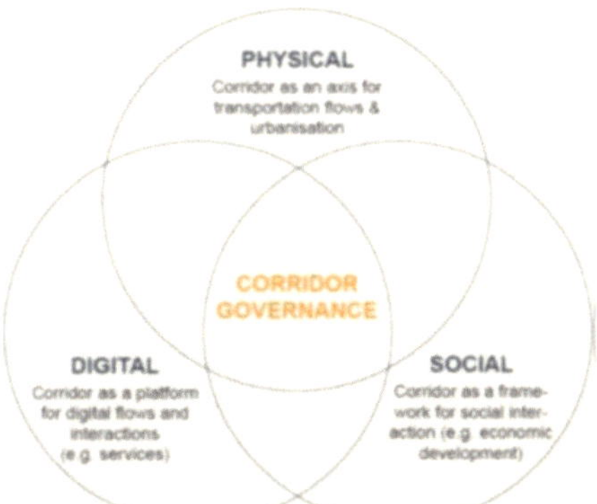

Figure 1.
Conceptual framework for broadening the perspective of corridor functionalities [14].

Dimension	Interactions/Flows
Physical	Customer Flows
	Daily Commuting
	Student Mobility
	Tourism & Leisure Flows
	Business Travel Flows
Social	Research Cooperation
	Trade Interaction
	Inter-Firm Cooperation
	Family/Social Interactions
	Student Interactions
	Planning Cooperation
	Purchase flows
Digital	Network traffic flows
	Monetary flows
	Communication flows

Table 1.
Categorization of example data sources for corridor development (shortened version) [14].

The Data-Based Territorial Management Strategy and the action plans are the first proposed measures through corridor cooperation, governance, and territorial development. One of the proposed models is also to establish and foster collaboration and partnerships [14].

3. The Iberian border territory along the border regions of Alto Alentejo (Portugal) and Extremadura (Badajoz-Spain)

Most of the cross-border region, Sparsely Populated Area (SPA), as well as Alto Alentejo (Portugal) and Extremadura-Badajoz (Spanish side of the border), are located at the margins of the administrative boundaries, small villages poorly connected by road infrastructure, weak urban network, low-density settlement, continued population losses and poor access to public transportation what isolates these settlements from the other surrounding areas.

In fact, SPAs can be schematized as a disorder of inadequately related local economies and communities, physically - i.e., secondary transport infrastructure, institutionally - i.e., loss of inter-municipal cooperation; or functionally - i.e., reduction of commuting beyond municipal borders. For example, in the Spanish case, constant SPAs can be observed along the borders of NUTS2 regions. Also, SPAs could attempt to 'reconnect' these populations across geographical borders to properly create a decisive mass of economic and social bonds from within, developing their collective ability to solve joint difficulties and unlock shared possibilities [13].

Contextually, **Table 2** represents of difficulty and obstacles in coordination and comprehensive border corridors governance based on modern studies (see: [10–13]), where have been and highlighted the common regional development between Portugal and Spain.

Common regional strategy	Challenge and obstacles in coordination and comprehensive border corridors governance: establishing integrated territorial development
Transnational coordination	Increasing number of new initiatives
	Setting up a territorial monitoring system
	Better macro-regional institutional cooperation
	Increasing number of regional stakeholders contributing to the implementation process
	• Lack of qualitative shifts of the political agendas
	Marketing and advertisement strategies are not coordinated
	The Euro-city is fertile to attract tourism and cultural activities
	Political - Strategic Several infrastructures are duplicated in nearby areas/ territories
	• Unemployment rates are still high in the territory
Political coordination and governance	Planning cooperation
	Integration of Common regional strategy objectives in domestic and European documents/projects, in particular regarding sectoral policies
	Number of strategic projects
	Continued population losses and poor access to public transportation
	• Several infrastructures are duplicated in nearby areas/territories
	• Sharing resources should be considered to avoid duplication
	• Lack of political commitment towards crossborder areas goals and EU standard's
	• Transparency and political commitment should be increased
	• Common Master-Plans and long-term objectives should be considered
Cross-sector coordination	GDP/ per capita values are considerable low
	Increasing horizontal coordination activities
	Unemployment rates are still high in the territory
	Data-Based Territorial Management Strategy
	Viewing the corridor as a framework for social interaction.
	Common activities and projects increasing number of actors, networks, meetings, projects
	Better tourism and Business Travel Flows
	The issue of connectivity can be related to the upgrading of communication and transport systems (CTS) to support the development of both extra-regional and intra-regional relations.
	Strengthen the evidence base for evidence-based planning in European growth corridors connecting to the TEN-T transport infrastructure network
Multi-level coordination	Participation of all levels
	Involvement of political stakeholders, e.g. the ministers meeting
	Increasing citizen engagement
	Increase transparency on multilevel governance structures and political governmental techniques.
	Improve institutional capacity, economic efficiency, competitiveness, inclusion and sustainability principles

Table 2.
Challenge and obstacles in coordination and comprehensive border corridors governance: establishing integrated territorial development.

4. Discussion and conclusions

In general, CBC planning reveals a loss of engagement at the local, regional, and meso-level, which need to be more actively included in the EU and macro-regional level. This is an extra obstacle to borderlands' successful regional development [9–11, 13]. Thus, it can also be assumed that at this level of CBC planning, there is a shortage of soft, resilient forms of governance which demands a tremendous political commitment as well as the use of already learned lessons regarding coordination and management at different levels that could be applied in this particular case study. The EU funds numerous projects and strategies on the Iberian Peninsula. However, in addition to all these projects and strategies along the Portuguese-Spanish border, several obstacles should be considered for further better territorial and sustainable development [10–22]. Some of them are presented in **Table 2**.

The development of a common regional strategy needs all members' involvement in the decision-making process to succeed all obstacles in coordination and comprehensive border corridors governance to establish integrated territorial development [1–3, 7, 8]. A case study of the CBC project in Iberian territory was presented to understand better the barriers to well-developed joint regional growth in Iberian territory. Several studies have already been conducted, which allowed us to analyze and evaluate outcomes. Border areas, which are also sparsely populated, should certainly support the EU in terms of flexible policies to enable their sustainable development.

Moreover, this study also aims to be useful for policymakers in moving from traditional data sources in the formation and planning of growth corridors to new, more comprehensive data sources that will give better results in preparing joint territorial development strategies in CBC areas [14].

Acknowledgements

The authors would like to acknowledge the financial support of the National FundsFundação para a Ciência e a Tecnologia, I.P. (Portuguese Foundation for Science and Technology) by the project UIDB/05064/2020 (VALORIZA – Research Centre for Endogenous Resource Valorization). Also, the authors wish to acknowledge funding for this research work from the VI Regional Research Plan and the Regional Government of Extremadura and the European Regional Development Fund (ERDF), associated with financing the research group Sustainable Development and Territorial Planning (GR18052), the Environmental Resources Analysis Research Group (GR18054) and Functional Study of Mediterranean Ecosystems (GR18078).

Conflict of interest

"The authors declare no conflict of interest."

Author details

Ana Vulevic[1,5]*, Rui Alexandre Castanho[2,3,4,5,6], José Manuel Naranjo Gómez[4,5,7], José Cabezas[5,8], Luis Fernández-Pozo[5,8], Jacinto Garrido Velarde[4,9], José Martín Gallardo[10], Sérgio Lousada[4,5,11] and Luís Loures[5,12]

1 Institute of Trasportation CIP, Belgrade, Serbia

2 Faculty of Applied Sciences, WSB University, Dąbrowa Górnicza, Poland

3 Aquageo Ambiente Legal—University of Campinas, Brazil (UNICAMP), Campinas, Brazil

4 CITUR - Madeira - Centre for Tourism Research, Development and Innovation, Funchal, Madeira, Portugal

5 VALORIZA - Research Centre for Endogenous Resource Valorization, Polytechnic Institute of Portalegre (IPP), Portalegre, Portugal

6 School of Business and Economics and CEEAplA, University of Azores, Ponta Delgada, Portugal

7 Agricultural School, University of Extremadura, Badajoz, Spain

8 ARAM, University of Extremadura, Badajoz, Spain

9 University of Extremadura, Badajoz, Spain

10 Functional Studies of Mediterranean Ecosystems, University of Extremadura, Badajoz, Spain

11 Faculty of Exact Sciences and Engineering (FCEE), Department of Civil Engineering and Geology (DECG), University of Madeira (UMa), Funchal, Portugal

12 Research Centre for Tourism, Sustainability and Well-being (CinTurs), University of Algarve, Faro, Portugal

*Address all correspondence to: anavukvu@gmail.com

References

[1] Van Houtum, H. An Overview of European geographical Research on Borders and border regions, J. Borderl. Stud. 2000, 15, 1.

[2] Decoville, A.; Durand, F.; Sohn, C.; Walther. O. Comparing cross-border metropolitan integration in Europe: towards a functional typology. J. Borderl. Stud. 2013, 28, 2, 221-237.

[3] Reitel, B. Border temporality and space integration in the European transborder agglomeration of Basel. J. Borderl. Stud. 2013, 28, 2, 239-256.

[4] United Nations (UN). UN Study: Transport for Sustainable Development - The case of Inland Transport. UNITED NATIONS New York and Geneva, 2015.

[5] Peyrony, J.; Denert, O. Planning for Cross-Border Territories: The Role Played. Raumforsch Raumordn 2012, 70, 229-240. Springer-Verlag, doi 10.1007/s13147-012-0156-0

[6] European Commission. DG Regio. Passenger rail accessibility in Europe's border areas. WP 11/2017, p. 3, 2017.

[7] Szary, A. L. 2012. Walls and Border Art: The Politics of Art Display. J. Borderl. Stud. 2012, 2, 213-228.2427.12071

[8] Decoville, A.; Durand. D. Building a cross border territorial strategy between four countries: wishful thinking? Eur. Plan. Stud. 2016, 24, 10.

[9] ESPON PROFECY (2018) Inner peripheries in Europe - Possible development strategies to overcome their marginalising effects. Policy Brief. Available at: https://www.espon.eu/inner-peripheries-brief.

[10] MEDEIROS, E. (2016). Territorial Impact Assessment and Public Policies: The case of Portugal and the EU. Public Policy. Port. J, 1, 51-61.

[11] CASTANHO, R., (2017). Sustainable Urban Planning in Transboundary Areas – Analysis of Critical Factors for Territorial Success. Ph.D. Thesis. University of Extremadura. Badajoz, Spain

[12] Vulevic, A., Castanho, R., Gómez, N., Loures, L., Cabezas, J., Fernández-Pozo, L., Gallardo, M. Accessibility Dynamics and Regional Cross-Border Cooperation (CBC) Perspectives in the Portuguese—Spanish Borderland, 2020 Sustainability 2020, *12*(5), 1978; doi:10.3390/su12051978

[13] ESPON BRIDGES Territories with Geographical Specificities, Applied research, Draft final report, 2018

[14] ESPON - Potentials of big data for integrated territorial policy development in the European growth corridors (Big Data & EGC), Targeted Analysis, Executive Summary, 28/06/2019

[15] Gløersen E (2012) Renewing the Theory and Practice of European Applied Territorial Research on Mountains, Islands and Sparsely Populated Areas. Regional Studies 46(4): 443-457. DOI: 10.1080/00343404.2012.665989.

[16] Giordano B and Dubois A (2018) Combining territory and competitiveness in EU Regional Policy? Analyzing ERDF investment profiles in regions with specific geographical features. Regional Studies 0(0): 1-10. DOI: 10.1080/00343404.2018.1495323.

[17] Castanho, R.A., Loures, L., Cabezas, J., and Garrido Velarde, J. (2020). The Socio-Cultural Factors in the Common Strategic Planning in European Territories: Principles

for Sustainable Development Extracted from Borderlands. WSEAS TRANSACTIONS on ENVIRONMENT and DEVELOPMENT, DOI: DOI: 10.37394/232015.2020.16.79

[18] Castanho, R.A. (2020). The Relevance of Political Engagement and Transparency in Cross-Border Cooperation (CBC) Environments. Analyzing Border cities in Europe. Lex localis - Journal of Local Self-Government, Vol. 18, No. 3, pp. 487-502, July 2020 https://doi.org/10.4335/18.3.487-502(2020).

[19] Jurado Almonte, J., Pazos-Garcia, F. and Castanho, R.A. (2020). Eurocities of the Iberian Borderland: A Second Generation of Border Cooperation Structures. An Analysis of Their Development Strategies. Sustainability 2020, 12(16), 6438; https://doi.org/10.3390/su12166438

[20] Castanho, R.A. (2019). Identifying Processes of Smart Planning, Governance and Management in European Border Cities. Learning from City-to-City Cooperation (C2C). Sustainability 2019, 11, 5476; doi:10.3390/su11195476.

[21] Castanho, R.A., (2019). Cross-Border Cooperation (CBC) the Paths to Achieve Territorial Success for EU Territories: the Critical Factors. European Week of Regions and Cities University 7th Master Class on EU Cohesion Policy. BOOK OF PAPERS 6th – 10th October 2019 - Brussels, Belgium #EURegionsWeek. pp.156-159.

[22] Loures, L., Castanho, R.A., Naranjo Gómez, J., Lousada, S., Fernández-Pozo, L., Cabezas, J., and Loures, A. (2019). Impactos Socioculturais da Cooperação Transfronteiriça (CT) no Espaço Europeu. Fronteiras: Journal of Social, Technological and Environmental Science http://periodicos.unievangelica.edu.br/fronteiras/ v.8, n.3, set.-dez. 2019 p. 292-312.

Section 3

Ultra-Peripheral Territories: The Islands

Chapter 4

Alignment between the Strategic Plans of Island Regions and the Agenda 2030 for Sustainable Development

Deolésio Mendes, Ana José and Joaquim Mourato

Abstract

The climate of increasing competitiveness between cities, leads to an urgent need for territorial actors to find solutions, so that cities become a more attractive space for different targets. Although cities are considered centers of innovation and social and economic development, their unplanned and unsustainable growth has resulted in harmful consequences for the environment, community, tourism, among others. This research aims to make a comparative analysis between the Strategic Plan for Sustainable Development of Cape Verde, the National Development Plan of São Tomé and Príncipe and the Proposal for the Sustainable Development Plan of the Autonomous Region of Príncipe, considering the theoretical contributions and the guidelines of the UN 2030 Agenda for Sustainable Development. It concludes about the similarities and differences between the plans, both in terms of the depth of the elements analyzed, as well as in alignment with the Agenda 2030.

Keywords: sustainable development, strategic planning, United Nations 2030 agenda, island regions

1. Introduction

We live in a climate of increasing competitiveness between cities, where the various territorial actors strive to find solutions, so that the territories become, in a more resilient, more attractive, more inclusive and with a better quality of life.

For this reason, this study aims to contribute to the understanding of the alignment of the Strategic Plan for Sustainable Development of Cape Verde, São Tomé and Príncipe and the Autonomous Region of Príncipe in compliance with the guidelines of the UN Agenda 2030 for Sustainable Development.

The Autonomous Region of Príncipe (ARP), in the context of small African island territories, has been standing out, thanks to small measures implemented in the conservation and preservation of nature and sustainable tourism, which guaranteed it the title of Reserve in 2012 World Organization for Education, Science and Culture (UNESCO) World Biosphere [1] and, recently, a praise from the United Nations (UN), as an example to follow.

It is imperative to plan and manage the future of the territories rationally, in this environment of uncertainty and ambiguity. This is because, sustainable development imposes on cities a "transparency of processes and mechanisms that, in a multidisciplinary and multisectoral way", requires the cooperation and collaboration of all its actors, so that the strategy developed reconciles with their interests [2, 3].

That said, it can be considered that a wide field of reflection is open on the main difficulties that small island states have been facing, in the face of the emergence of new paradigms imposed by globalization.

2. Literature review

2.1 Sustainable cities as a development strategy

Considering globalization and the opportunity cost of living in cities in relation to the rural environment, it appears that cities have been registering a significant increase in population [4]. Although these are considered to be the center of innovation and social and economic development, their unplanned and often unsustainable growth has resulted in harmful consequences for the environment, such as: soil degradation, social inequality, pollution, quality of life, among others [5].

In this sense, the territorial actors felt the need to develop and implement strategies, which enhance the promotion and territorial valorization, strive for environmental preservation, favor social responsibility and provide, above all, the satisfaction of the needs of all stakeholders [5–7].

It is in this context that sustainable development, which is characterized by the ability to meet the needs and challenges of the present, without ever compromising or neglecting the needs of future generations, assumes a leading role in the development of sustainable cities [8].

In this line of thought, the sustainable development of cities and territories must be "a process of transformation, in which the exploitation of resources, the direction of investments, the orientation of technological development and institutional change are harmonized and reinforce the present potential and future in order to meet human needs and aspirations" [8].

For this reason, the sustainable development of cities must be based on three dimensions (environmental, economic and social), in order to facilitate the balance between man and nature. Government institutions have been adopting policies for sustainable development of cities, with the objective of safeguarding the interests and resources of the territories, so that they are always in competitive equality in relation to other cities, thus gaining more and more attractive capacity before stakeholders [9, 10]. This type of development must motivate a committed attitude among all territorial agents, in order to guarantee competitiveness and social cohesion [11, 12].

In addition, cities must be aware of the challenges inherent in their development, in order to build and envision the future, encouraging the use of innovative and creative skills, guided by a scientific and technological research and development (R&D) culture and investing in following views [5, 13–15]:

- *Prosperity*: offering high levels of quality of life to qualified people, in order to foster the knowledge that generates innovation and creativity and, consequently, attract and fix resources, especially financial;

- *Healthy*: promote the proximity between man and nature, through the valorization of natural resources and assume the coefficients and the reduction of the ecological and carbon footprint, through healthy spaces;

- *Resilience*: establish a posture committed to future generations, conserving the heritage and, using current trends, develop measures to bridge emerging paradigms, always safeguarding endogenous resources;

- *Justice*: not to deprive its citizens of the use of citizenship and their cultural identity;

- Inclusion: foster social cohesion and inclusion of individuals, giving them opportunities for education, health, security and justice;

- *Connection*: making connection with the world viable, through cooperation, complementarity and sustainable development.

2.2 Agenda 2030 guidelines for sustainable development

The leaders of each country and the UN, developed certain measures aimed at, "ending poverty and hunger everywhere; combat inequalities within and between countries; build peaceful, just and inclusive societies; protect human rights and promote gender equality and the empowerment of women and girls; and to ensure the lasting protection of the planet and its natural resources (...) create conditions for sustainable, inclusive and economically sustained growth, shared prosperity and decent work for all, taking into account the different levels of development and national capacities" [16].

In this sense, we sought to develop measures that were related to essential elements for sustainable development, being: people, planet, prosperity, peace and cooperation [16], as detailed:

- *People*: Eradicate poverty and strive for equality and equity among human beings, on a healthy planet;

- *Planet*: Protect it against the harmful actions of human beings through development policies and sustainable management of resources;

- *Prosperity*: Ensure that economic, social and technological development, proceed in "harmony with nature";

- *Peace*: "Promote peaceful, fair and inclusive societies that are free from fear and violence";

- *Cooperation*: combining all existing means and synergies, so that the implementation of this Agenda is based on a revitalized cooperation/partnership on a global scale, that is, "on the basis of a spirit of reinforced global solidarity, focused especially on the needs of the poorest and most vulnerable and with the participation of all countries, all stakeholders and all people".

In this sense, through the Agenda 2030 for development, a set of 17 Sustainable Development Goals (SDGs) (**Figure 1**), 169 goals and 230 indicators was established, as a way to achieve sustainable development for the planet [16].

However, the implementation and consequent expected results of the UN Agenda 2030 SDGs for sustainable development and the respective guidelines, will only be possible if, on the one hand, we seriously strive for a culture of cooperation at all levels between countries, organizations, civil societies and all actors interested in the sustainable development of a territory and, on the other, if each country is

Source: [16]

Figure 1.
The UN's sustainable development goals. Source: [16].

responsible for its own economic and social development, regardless of its political leadership, provided that its policies are awake and compatible with the international rules, to eradicate poverty and spread sustainable development.

3. Methodology

The study that is presented is of a qualitative nature in which it allows to relate the concept with the phenomenon, through the perception of the existing dynamics between the method of induction and/or subjectivity [17]. In other words, studies that choose a qualitative approach, assume that phenomena need meaning. For this reason, the researcher needs to systematize the context, not only using subjectivity, but also the "theoretical-methodological resources that make it possible to perform such a task", in order to find "a path that will lead him to the most reliable interpretation possible" [18] referenced by [19].

For a better understanding and ease of analysis of strategic plans, the content analysis methodology was used, which is characterized by a "set of communication analysis techniques, which uses systematic and objective procedures for describing the content of messages" [17].

This methodology involves three phases: pre-analysis, material exploration and data processing and interpretation [20]. In this specific case, they corresponded to:

Pre-analysis - reading, selection of strategic plans and constitution of bodies, which concerns the set of three strategic plans;

Material exploration - refers to the creation of a document analysis instrument;

Treatment of results and interpretations - equates to analysis and interpretation of the Cape Verde, São Tomé and Príncipe and ARP plans.

For a better understanding of the analysis that took place in each strategic plan and consecutively the alignment that took place between them and the UN Agenda 2030 for sustainable development, a document analysis instrument was adopted, which focused essentially on the sustainability parameters, since it sought to evaluate the following elements: Stakeholder involvement; Economic prosperity; Inclusion and social cohesion; Social justice; Connectivity with the world;

Approximation between man and nature; Territorial resilience; Innovation and creativity; Scientific and technological R&D.

For each of these elements, a scale of 0 to 5 was used, which allowed to assess the level of depth, of each element in the alignment and in compliance with the guidelines of the UN Agenda 2030. As a result, it was also possible to evaluate the depth index for each strategic plan, taking into account the level of depth assigned.

Regarding the connotation of the scale, it is important to note that the values correspond to the following: 0- Does not exist; 1- Very weak; 2- Weak; 3- Reasonable; 4- Strong; 5- Very strong.

It is important to note that the different scores attributed to each element resulted from the content analysis carried out on the strategic plans.

With regard to the depth index, this is an indicator that will allow a more quantitative assessment of the extent to which the different strategic plans comply with the guidelines of the UN Agenda 2030. In other words, this index reflects the level of contribution of each plan to the fulfillment of the UN Agenda2030 guidelines for sustainable development. For the sum of contribution of each SDG (score) will be the overall depth index of the plan. It is important to note that the scores refer to the level of depth that is attributed to each SDG in the different plans.

The way to calculate the depth indices will be as follows:

$$\text{Depth Index (SDG)} = \frac{Score}{(5\times 19)}\, x\ 100 \tag{1}$$

$$\text{Total Plan Depth Index} = \frac{\sum \text{Scores}}{(5\, x 19)}\, x\ 100 \tag{2}$$

It should be noted that in the formula, 5 represents the maximum depth level that can be attributed to each element and 19 is related to the number of elements susceptible to evaluation.

Therefore, it is important to clarify that the instrument for aligning strategic plans in compliance with the UN Agenda 2030, was composed of the SDGs illustrated in **Figure 1**, the Depth Levels and the Depth Index.

4. Results and discussion

4.1 Strategic plan for sustainable development of Cape Verde

The main challenges to overcome for the development and progress of the archipelago, focuses on:

Combat vulnerabilities (economic, environmental and social) and apply measures that make the country more resilient - betting on standards that aim to make the islands more inclusive, with a sustainable base economy, in order to challenge the "*external dependence, the unemployment (at all levels), poverty, inequality in the distribution of income, reduced opportunities for emigration and, consequently, a drop in remittances*" (...) and guarantee the valorization and use of endogenous products;

Bet on the valorization of the islands through the decentralization of power, and also in the valorization of the endogenous opportunities and resources;

Striving for education of excellence - ensuring the adjustment, integration, coherence, consistency of preschool to higher education by introducing them to the main information and communication technologies, in order to train teachers and students for the new form of development. In parallel with this, adapt training offers "*to the needs of the digital and nanotechnological economy*" and train human resources in order to ensure sustainable development;

Streamline the integration of international development agendas, namely Agenda 2030 and Agenda 2063 of the UN and the Accelerated Modality of Action of Small Island Developing States (SIGS) - Samoa Pathway.

The following vision for Cape Verde of the future is defined "*A developed, inclusive, democratic Cape Verde, open to the world, modern, secure, where full employment and full freedom prevail*".

In order to materialize this vision, it became essential to set four strategic objectives, namely:

"*Making Cape Verde an Economy of Circulation Located in the Middle Atlantic*" - It implies betting on "*Creation of a logistical port to supply international ships that pass through Cape Verde*"; "*Creation of a logistics airport for the international distribution of passengers and cargo*"; Attraction of Companies and Investments and local business promotion; "*Creation of an international financial platform*"; Foster the participation of the migrant population in the development process; Bet on Tourism; "*Development of the Digital and Nano-technological Economy*";

"*Ensuring Economic and Environmental Sustainability*";

"*Ensuring Social Inclusion and Reducing Social and Regional Inequalities and Asymmetries*";

"*Strengthening Sovereignty, Valuing Democracy and Orienting Diplomacy to the Country's Development Challenges*".

For Cape Verde to achieve these objectives, it will need to maintain good relations with all its stakeholders, that is, having a good relationship with non-governmental organizations, the public and private sector, among others, in order to guarantee support, namely the Green Climate Fund, Global Fund for the Environment, support from its UN experts, to instill in the Cape Verdean population good practices of sustainability and use of endogenous resources, among others.

Table 1 illustrates the assessment of sustainability parameters in Cape Verde's strategic plan.

Through this illustration it can be said that in the Cape Verde Strategic Plan of Strategic Development (SPSD), sustainability criteria are in general strongly valued, since they present a level of depth, mostly strong, except for the involvement of stakeholders, connectivity with the world and territorial resilience, which have a very strong level of depth. Alongside this, it is necessary to point out that the Government of Cape Verde is aware that a large part of the financing to support this SPSD will depend on the "external financing of banks and development funds, in order to reinforce the cooperation ties between the institutions".

Table 2 illustrates the alignment of SPSD in compliance with the UN Agenda 2030 guidelines.

It is concluded that, in general, SPSD contributes positively to the fulfillment of the UN Agenda 2030 SDGs, presenting a depth index of 75.79%. Nevertheless, there are some SDGs that need greater depth to globally comply with the UN Agenda 2030 guidelines, such as: eradicating hunger, protecting marine life, gender equality and climate action. Therefore, improving performance in these SDGs leads to a more robust alignment of SPSD with the guidelines of the UN Agenda 2030.

4.2 National development plan of São Tomé and Príncipe

For São Tomé and Príncipe (STP) to achieve its development, it must overcome the following challenges:

Dimensions	Elements	Level of depth					
		Does not exist	Very Weak	Weak	Reasonable	Strong	Very strong
Sustainability	*Stakeholders* involvement						x
	Economic Prosperity					X	
	Social inclusion and cohesion					X	
	Social justice					X	
	Connectivity with the world						x
	Bringing man closer to nature					X	
	Territorial resilience						x
	Innovation and creativity					X	
	Scientific and Technological I&D					X	

Table 1.
SPSD evaluation.

Develop the "productive base" of the country's economy and promote the "diversification of sources of growth and employment", so that youth and women unemployment and can be mitigated, and also supply the local market with endogenous products, in order to "alleviate" the export rate;

To train human resources in public administration, using "methods", "instruments and working conditions", in order to facilitate the implementation and monitoring of development policies, since the political interest seriously influences the country's development, especially in the social and economic scope;

Promote gender equity and equality throughout the development process and especially in public administration;

Generate and "modernize the infrastructure network" to "support development (energy, roads, ports, airports, water)" and Mitigate territorial vulnerabilities (environmental and social).

The vision that we intend to have focuses on: "São Tomé and Príncipe 2030: The Country we want to build", the objective of this vision is to build a country so that "São Tomé and Príncipe can live decently in a stable, egalitarian, democratic and solidary, in the process of modernization and offering quality services at regional and global level, without any form of discrimination against women".

Regarding the strategic objectives, the following were outlined for STP:

In the economic sphere - "Improving the exploitation of the country's development potential, accelerating economic growth and striving to integrate into the regional and global economy", through "diversifying the economy and expanding its production base", "increasing production and diversifying of food culture and expansion of exploitation culture", "optimization of the exploitation

ODS	Level of depth						Score	Depth of index (%)
	0	1	2	3	4	5		
Eradicate poverty					x		4	4,21
Eradicate hunger			x				2	2,11
Quality health					x		4	4,21
Quality education					x		4	4,21
Gender equality				x			3	3,16
Clean water and Sanitation					x		4	4,21
Renewable and affordable energy					x		4	4,21
Decent work and economic growth					x		4	4,21
Industry, Innovation and Infrastructure					x		4	4,21
Reduce Inequalities					x		4	4,21
Sustainable cities and communities					x		4	4,21
Sustainable production and consumption					x		4	4,21
Climate action				x			3	3,16
Protect marine life			x				2	2,11
Protect life on earth					x		4	4,21
Peace, justice and effective institutions						x	5	5,26
Partnerships to implement the objectives						x	5	5,26
Means of implementation					x		4	4,21
Means of monitoring					x		4	4,21
Total						**68**	75,79	

Table 2.
SPSD alignment in compliance with UN guidelines.

of fisheries resources", to boost services, development of tourism and small industries;

In the social sphere - "Accelerate and deepen reforms to significantly improve the human development index of São Tomé and make substantial progress towards the conservation of the SDGs". For this to become necessary, "strengthening human capital and governance", improving health services, "promoting youth, consolidating families and protecting vulnerable groups" and "valuing national culture, supporting the development of sport and inclusion of the diaspora";

In the institutional/governance domain - "Improving the capacity for strategic management of national development, strengthening good governance and democracy", through "improving the quality, speed and accessibility of justice", fighting crime, "developing the financial sector", "Improving strategic development management", "improving public finance management", "strengthening governance", "gender institutionalization", "strengthening local development poles and promoting decentralization" and "consolidating international cooperation and preserving national sovereignty";

In the field of infrastructures - "development of economic and social infrastructures" and "promotion of the development of telecommunications" and information and communication technologies;

In the field of the environment - "Improving land management and preserving the environment".

Given this scenario, the STP established the following target audiences: tourists, resident population and investors.

Maintaining a good relationship with its partners becomes an "agenda" phrase for the Democratic Republic of STP, since about 89% of its budget is financed by external partners.

Using the document analysis instrument, the National Development Plan for São Tomé and Príncipe (NDPSTP) is evaluated as follows (**Table 3**).

Analyzing **Table 3**, it was found that there is a need to deepen, globally, even more in the elements of sustainability in the NDPSTP, since most of these elements have a reasonable depth index, except the elements Stakeholder involvement, economic prosperity and Inclusion and social cohesion, with a strong or very strong level of depth.

The implementation of the NDPSTP is dependent on a set of reforms, which aim to improve the strategic management of STP development in compliance with the UN Agenda 2030 guidelines:

- "Operationalization of new planning systems (…) through the implementation of a true management based on results";
- "Accelerate the implementation of the public finance reform action plan";
- "The development of a national action plan for the implementation of the Paris Declaration on the effectiveness of Public Development Assistance and the Strengthening of Coordination with Technical and Financial Partners";
- "Consolidation of an information system to respond to information needs".

Table 4 evaluates the alignment of the NDPSTP with the aforementioned UN Agenda 2030 guidelines.

Considering **Table 4**, the NDPSTP presents a depth index of 65.26% in compliance with the UN Agenda 2030 guidelines. This situation must be improved, focusing on deepening the following areas: eradicating hunger and poverty, sustainable cities and communities and protecting marine life. Other SDGs have an average depth level, and there is still a very large margin for progression.

4.3 Strategic plan for sustainable development of the autonomous region of Príncipe - Príncipe 2030

The Autonomous Region of Príncipe must address the following constraints in order to achieve its development:

People - Low number of teachers at all levels, Insufficient offer of professional training opportunities, poor coverage of the health system, limited social protection policy and gender equality policies, difficulty in preserving material and immaterial heritage, dependence on the community International;

Dimensions	Elements	Level of depth					
		Does not exist	Very Weak	Weak	Reasonable	Strong	Very strong
Sustainability	*Stakeholders* involvement					x	
	Economic Prosperity						x
	Social inclusion and cohesion					x	
	Social justice				x		
	Connectivity with the world				x		
	Bringing man closer to nature				x		
	Territorial resilience				x		
	Innovation and creativity				x		
	Scientific and Technological I&D			x			

Table 3.
Evaluation of the NDPSTP.

Planet - Use of non-renewable ("dirty") energy sources, limited access to drinking water and sanitation, lack of public engagement in protecting and conserving the environment and good practices, inadequate waste collection and deposition and reduced recycling, use of products harmful to the environment, uncontrolled exploitation of natural resource areas, negative impact of climate change;

Prosperity - Persistence of non-responsible tourism investments, lack of promotion of entrepreneurship and innovation, deficient airport infrastructure, lack of investment in renewable energies, limited training of tourist agents, inadequate employment conditions, limited digitalization of the island;

Peace and Partnership - Lack of proximity policies for law enforcement officers to communities, perfectable justice system, limited promotion of Public-Private Partnerships, lack of training and continuous training, difficult mobilization of financial resources for investment, lack of harmonization of financing in order to avoid duplication of actions;

Planning - Deficient accountability system and associated regulatory environment, insufficient understanding of responsibilities on the part of institutions, limited institutional articulation, little understanding of the mechanisms for validating, implementing, monitoring and evaluating plans and projects, lack of regional cartography and registration, failure to integrate the profile of the regional demographic dividend in the planning process.

The vision that we intend to have on Príncipe in 2030, is as follows: "Until 2030, Príncipe Island, a global biosphere reserve, is an international reference for biodiversity conservation and sustainable development, inclusive and resilient to climate change, for through conciliation between responsible tourism and the green

ODS	Level of depth						Score	Depth of index (%)
	0	1	2	3	4	5		
Eradicate poverty				x			3	3,16
Eradicate hunger			x				2	2,11
Quality health					x		4	4,21
Quality education				x			3	3,16
Gender equality					x		4	4,21
Clean water and Sanitation				x			3	3,16
Renewable and affordable energy				x			3	3,16
Decent work and economic growth					x		4	4,21
Industry, Innovation and Infrastructure				x			3	3,16
Reduce Inequalities					x		4	4,21
Sustainable cities and communities			x				2	2,11
Sustainable production and consumption					x		4	4,21
Climate action				x			3	3,16
Protect marine life	x						0	0,00
Protect life on earth					x		4	4,21
Peace, justice and effective institutions					x		5	5,26
Partnerships to implement the objectives					x		4	4,21
Means of implementation				x			3	3,16
Means of monitoring					x		4	4,21
Total							**62**	**65,26**

Table 4.
Alignment of NDPSTP in compliance with UN guidelines.

and blue economy, guaranteeing people's quality of life ", based on the following values:" Sustainability, Inclusion, Resilience, Innovation, Efficiency, Participation, Transparency and Justice ".

Of the seventeen SDGs stipulated by the UN Agenda 2030, ARP established ten of these strategic objectives as a priority, to be achieved in a period of 10 years, materializing in:

Environmental Sustainability - Protect, restore and promote the sustainable and responsible management and use of land and marine resources. Stop and reverse the degradation of ecosystems and the loss of biodiversity;

Resilience to Climate Change - Adopt effective and efficient mitigation and adaptation measures to combat climate change and its impacts;

Responsible Tourism - Promote responsible and sustainable tourism, which generates decent jobs and economic diversification. Valuing local culture and products, with a positive impact on the environment, society and the prosperity of the island;

Green and Blue Economy - Promote responsible and sustainable investments and entrepreneurship, with a special focus on organic farming, agroforestry and fishing, enhancing decent employment and local value chains.

Quality of Life for All - Ensure social inclusion, security and universal health coverage, promoting equitable opportunities;

Resilient Infrastructures - Develop collective, quality, reliable, sustainable, intelligent, integrated and resilient infrastructures and equipment that promote social well-being, economic development and better connectivity between the island and the world;

Innovation and Funding - Foster innovation and ensure inclusive and quality education, training local human resources to ensure personal fulfillment and the sustainable development of the island;

Participatory Planning - Implement participatory policies, plans, legal and monitoring frameworks for sustainable and inclusive socio-economic and territorial development;

Transparent Governance - Ensure participatory, efficient, open and transparent governance that promotes access to information and legal certainty;

Financing for Development - Mobilize financial resources through national, international and public-private partnerships. Generate endogenous revenues and ensure transparent and efficient financial management.

With regard to sustainability parameters, the Príncipe plan was assessed as follows (**Table 5**).

Taking into account **Table 5**, the Príncipe Plan 2030 presents some weaknesses in the sustainability criteria, since some elements need greater depth, namely: Economic prosperity, Social justice, scientific and technological R&D and Connectivity with the world.

Dimensions	**Elements**	**Level of depth**					
		Doesn't exist	**Very Weak**	**Weak**	**Reasonable**	**Strong**	**Very strong**
Sustainability	*Stakeholders* involvement						x
	Economic Prosperity			x			
	Social inclusion and cohesion					x	
	Social justice			x			
	Connectivity with the world				x		
	Bringing man closer to nature					x	
	Territorial resilience					x	
	Innovation and creativity					x	
	Scientific and Technological I&D			x			

Table 5.
Evaluation of Príncipe plan 2030.

ODS	Level of depth						Score	Depth of index (%)
	0	1	2	3	4	5		
Eradicate poverty					x		4	4,21
Eradicate hunger				x			3	3,16
Quality health					x		4	4,21
Quality education					x		4	4,21
Gender equality					x		4	4,21
Clean water and Sanitation					x		4	4,21
Renewable and affordable energy				x			3	3,16
Decent work and economic growth			x				2	2,11
Industry, Innovation and Infrastructure				x			3	3,16
Reduce Inequalities				x			3	3,16
Sustainable cities and communities					x		4	4,21
Sustainable production and consumption				x			3	3,16
Climate action					x		4	4,21
Protect marine life					x		4	4,21
Protect life on earth					x		4	4,21
Peace, justice and effective institutions			x				2	2,11
Partnerships to implement the objectives					x		4	4,21
Means of implementation					x		4	4,21
Means of monitoring				x			3	3,16
Total							**66**	**69,47**

Table 6.
Alignment of the prince plan 2030 with UN guidelines.

The alignment of the Príncipe Plan 2030 in compliance with the UN Agenda 2030 is assessed as follows (**Table 6**).

It was found that the Príncipe Plan 2030 has a depth index of 69.47% in compliance with the UN Agenda 2030 guidelines. This index was negatively influenced, essentially, by the lack of depth in the areas "Decent work and economic growth" and "Peace, justice and effective institutions. It is also noted that no SDGs earned the maximum score, which means that there is a high margin for progression.

5. Conclusions

From the comparative analysis between the strategic plans for sustainable development of Cape Verde, Santo Tomé and Príncipe and Ilha do Príncipe, it was possible to determine similarities and differences.

The main similarities between the plans correspond to participation and inclusion, a focus on tourism, good governance and good relations with development partners, the need to attract investment and measures of resilience and sustainability (environmental, social and economic).

Conversely, the plans show some differences, especially in areas that need more depth. The Cape Verde plan emphasizes the need to deepen institutional capacity, the institutional relationship with the outside world, define ways to communicate its attractions, boost its business fabric, invest in attracting qualified personnel abroad and determine ways to promote the territory.

In the case of the São Tomé and Príncipe plan, the weaknesses revealed were in the following areas: bet only on the qualification of public administration personnel and not on basic education; develop concrete policies to attract staff trained abroad and promote a wider range of emigrants in the development strategy, establish strategies for territorial promotion, not only bet on basic strategies, create measures to combat illiteracy and encourage adolescents to complete compulsory education, among others.

Likewise, the Príncipe Plan 2030 presents a set of points that need to be improved, namely the territorial promotion strategies as a brand, define more consistent attractions given the market's competitiveness, promote entrepreneurship (social and business) and establish an instrument to monitor the Príncipe Plan 2030.

In addition, there were also some differences between the plans when assessing their alignment with the UN guidelines for sustainable development. It was found, through the depth index, that the Cape Verde plan is in line with 75.79% of the UN Agenda 2030 guidelines, while the São Tomé and Príncipe plan has a 65.26% index, and the Príncipe Plan 2030 the rate of 69.47%.

The depth index allows us to conclude that all the plans analyzed are positively aligned with the Sustainable Development Goals of the Agenda 2030, while identifying the elements that still have a wide margin for improvement.

In short, in order to enhance the sustainable development of these regions, it is recommended that the plans reinforce their performance, focusing in particular on the identified weaknesses.

Acknowledgements

Authors are grateful for the financial support from the Foundation for Science and Technology, I.P. (Portuguese Foundation for Science and Technology) by the project UIDB/05064/2020 (VALORIZA - Research Center for Endogenous Resource Valorization).

Author details

Deolésio Mendes[1], Ana José[2] and Joaquim Mourato[3,4]*

1 Polytechnic Institute of Portalegre, Portalegre, Portugal

2 Polytechnic Institute of Portalegre, C3i - Coordenação Interdisciplinar para a Investigação e Inovação, Portalegre, Portugal

3 Centro de Investigação para Valorização dos Recursos Endógenos (VALORIZA-IPP), Polytechnic Institute of Portalegre, Portalegre, Portugal

4 Centro de Investigação e Estudos de Sociologia (CIES-IUL), Lisboa, Portugal

*Address all correspondence to: amourato@ipportalegre.pt

References

[1] Camões Instituto da Cooperação e da Língua Portuguesa, 23-05-2013. [Internet]. Available from: https://www.instituto camoes.pt/sobre/comunicacao/noticias/ilha-do-principe-reserva-mundial-da-biosfera-apresenta-video. [Accessed: 2020-06-06].

[2] Barros C. Planeamento estratégico de marketing territorial e perspectivas de desenvolvimento na Região Autónoma da Madeira. Coimbra: Faculdade de Letras da Universidade de Coimbra; 2011

[3] Plano de Ordenamento do Espaço Marítimo. vol. Vol. 1. Lisboa: Ministério Do Ambiente e do Ordenamento do Território; 2010

[4] Lopes, A. O Programa Cidade Sustentável, seus Indicadores e Metas: Instrumentos Metodológicos para a Avaliação da Sustentabilidade no Município de Prata/MG Mina Gerais – Uberlândia: Faculdade de Arquitetura e Urbanismo e Design da Universidade Federal de Uberlândia, 2016.

[5] Euromonitor [Internet]. Available from: https://www.euromonitor.com/. [Accessed: 2020-01-15].

[6] Gaio S, Gouveia L. O Branding Territorial: uma abordagem mercadológica à Cidade. Revista A Obra Nasce. 2007:27-36

[7] Salgado M, Leitão M. Posicionamento da cidade de Coimbra face a potenciais concorrentes. Revista Turismo e Desenvolvimento. 2010;**14**:287-298

[8] Report of the World Commission on Environment and Development - Our Common Future. 1987.

[9] Década da Educação das Nações Unidas para um Desenvolvimento Sustentável: 2005-2014. Brasília, 2005.

[10] Gutierrez, M. O Marketing Territorial para o desenvolvimento sustentável da Euro-região Galiza-Norte de Portugal- Braga: Escola de Economia e Gestão da Universidade do Minho, 2013.

[11] Almeida A. Os Desafios das Cidades - Contributos da Gestão, Marketing e Planeamento Estratégicos, P. -. P. R. d. Silva: Ed.; 2013

[12] Bittencourt B. Políticas de Desenvolvimento Local Sustentável e o Terceiro Setor: Estudo de Caso na Região de Aveiro. Tese de Doutoramento, Lisboa: Instituto Superiro de economia e Gestão da Universidade de Lisboa; 2014

[13] Mendes JO. Futuro das cidades. Coimbra: Edições. Minerva. 2012

[14] Vilares ER, Magalhães F, Esteves M, e Tavares N, Cidades Sustentáveis M. Lisboa: Direção-Geral do Território. Ministério do Ambiente: Ordenamento do Território e Energia; 2020. p. 2015

[15] Brito C, Lencastre P. Novos Horizontes do Marketing. Alfragide: Publicações Dom Quixote; 2014

[16] Organização da Nações Unidades, 2015. [Internet]. Available from: https://unric.org/pt/. [Accessed: 2020-02-21].

[17] Bardin L. Análise de Conteúdo. Lisboa: Edições. 2006;**70**

[18] Alves-Mazzotti A, Geawandsznajder F. O Método nas Ciências Naturais e Sociais: Pesquisa Quantitativa e Qualitativa. Pioneira: São Paulo; 2000

[19] Barbosa, W. Projeto Veredas: Vozes Docentes Sobre a Prática Reflexiva, Rio de Janeiro: Juiz de Fora: Faculdade de Educação da Universidade Federal de Juiz de Fora, 2006.

[20] Bardin LA d CL. Edições 70. In: Lda. 1997

Chapter 5

The Azores Archipelago as a Region with Vast Potential for the Development of Adventure and Slow Tourism

Pedro Pimentel, André Oliveira, Gualter Couto, João Crispim Ponte and Rui Castanho

Abstract

WTO (World Tourism Organization) recognizes that adventure tourism is a journey that includes at least two of three elements: (a) physical activity, (b) natural environment; and (c) cultural immersion. With this in mind, the Autonomous Region of Azores, classified as one of the OR (Outermost Regions) by the European Union (EU), meets several of those elements. Therefore, the Archipelago shows several potential for the development of this typology of tourism. However, this insular region faces many challenges. Some of those obstacles are also great opportunities for regional development towards new sustainable models. For example, there is the islands' geographical location, which enables exceptional territorial development opportunities and growth opportunities. Contextually, the Archipelago's nine islands afford great variety to the destination, and the tranquil rural landscape fits the perfect framework for unparalleled, charming, and relaxing experiences. Moreover, the Archipelago's specific characteristics, as its volcanic nature, make the destination a perfect place to be explored. These are just some examples of the region's potential to develop this typology of tourism, based on nature, and consequently, a vehicle to achieve the so-desired regional sustainability.

Keywords: nature-based tourism, regional studies, strategic planning, sustainable tourism, ultra-peripheral territories

1. Introduction (Slow adventure tourism)

Adventure tourism does not present itself to a straightforward interpretation. It overlays with various distinct tourism sections and has unique characteristics that produce a context for unforgettable experiences. The description of adventure tourism offered by the Adventure Travel Trade Association (ATTA) has been embraced by the World Tourism Organization [1], who consider it as a trip that involves at least two of the following three components: physical activity, natural environment, and cultural immersion [1]. In turn, Buckley [2, 3] states that: "(...) *adventure tourism broadly means guided commercial tours where the principal attraction is an outdoor activity, which relies on features of the natural terrain and requires specialized sporting or similar equipment, and is exciting for the tour clients".* Therefore, slow adventure is different.

In this regard, Varley and Semple [4] introduced the concept of slow adventure, which they consider: "(...) *a celebration of the (ir)rationality of uncertainty, unpredictability, transience, experiment, and the emotional content of human experience, particularly in the context of the great outdoors and engagement with the "more-than-human world."* According to Varley and Semple [4] and Varley, Farkic, and Carnicelli [5], the concept of slow adventure is based on "(...) *an appreciation of the journey as an experiential dimension rather than the chore of getting to a destination".* Besides, the authors add that the concept of "slow" adventure is rooted in the Nordic conception of "*friluftsliv*" as the primary and simple activity of just being, or dwelling, in nature for long periods of time, which allows for the generation of rich experiences, a deep recognition of and spiritual immersion in the natural environment through involving in simple outdoor activities.

In 2015, the SAINT project, co-financed by the Northern Periphery and Arctic Programme, kicked off. This project focus was primarily to work with micro-businesses to improve their promotion of slow adventure activities and extend their marketing reach to new overseas and domestic consumers, leading to positive direct impacts in terms of more bed nights, extending visitors' stays and potentially stretching the seasons [6]. The project involved a significant partnership, including research centers, local governments, and SMEs from Scotland, Ireland, Finland, Norway, Iceland, Northern Ireland, and Sweden.

In the SAINT Project, "slow adventure"was defined as a form of tourism that avoids the quick-fix adrenalin-pumping hits of convenient adventure experiences in favor of slow, immersive journeys, living in and traveling through wild places and natural spaces—experiencing nature in its timeframe, seasons, weathers and variations (**Figures 1** and **2**) [6]. The project also defends that slow adventure activities draw upon ideas framed around the partners' cultures and histories, such as hunting, fishing and wild foods, love of and respect for nature and wildlife, and the relationship between food and the land/sea [6].

Furthermore, Varley and Semple [4] consider slow adventure as mainly related to outdoor living and journeying experience. They claim that there are four critical elements to this concept—time, nature, passage, and comfort.

Time: It is inevitably a crucial experiential component. The awareness of time passing during outdoor journeys is felt during the "passage" of the journey itself and natural change such as light and dark, tides, and weather (Varley and Semple, 2015). Time is manifested in a natural awareness of its passing during outdoor journeys [5].

Figure 1.
SAINT project marketing image.

Figure 2.
SAINT project slow adventure logo.

Nature: The effects of "nature" are acute in slow adventures due to the extended time of exposure. This direct engagement with natural forces insists that participants envelop themselves in their environment and surrender to it even (Varley and Semple, 2015). Nature refers to the natural setting and our effortful access to it [5].

Passage: The term "passage" refers not only to the physical journey through a physical landscape (as opposed to the passage over the landscape of the passenger), but it is also a journey of change and transformation, which takes time [4]. Passage, both physical and spiritual, is the navigation through time, space, and the self [5].

Comfort: This equally has several meanings in the context of a slow adventure. First, there is becoming comfortable with the challenges presented by the journey (sustained effort, for example). Also, comfort may be derived from a reconnection with the place, tradition, and history (linked to time). The traditional, rural life, imagined as slow, rich, and meaningful, can become a "refuge" landscape, a place to escape to [4]. Comfort implies being at ease with unusual challenges throughout the journey and might include reconnection with the place, and even with ourselves and others [5].

These four elements have in themselves the features that allow the slow adventure to create value to destinations: 1) more time dedicated to experiences will result in higher tourists' average stay, which is one performance indicator that the Azores need to improve on; 2) respectful, meaningful and responsible interaction with nature is of utmost importance to the Azores since nature-based tourism is its priority product; 3) passage is something that is felt and imprints itself in the participant, making places and events memorable, which can be magnified by the great landscapes and paradise-like places in the Azores; and 4) the deep connection with inspiring outdoors and small rural communities, which are the core features of the Azores, provide comfort and create the potential for repeat visits [6].

Based on the above-mentioned Slow Adventure was formally registered as a brand within the SAINT Project. Furthermore, the concept behind that brand is the most important legacy of the project, which can be incorporated into other destinations and products. The "Slow Adventure Guidelines," one of the project's outputs, refers to the fact that slow Adventure is all about creating a lasting impression and a sense of purpose for the consumer. These products are highly suited to tourism development in un-spoilt rural and peripheral areas. Therefore, recalling the implicit and respectful interaction with nature and communities, this concept goes hand-in-hand with sustainable tourism development. Moreover, it has a vast potential to reduce seasonality, increase the power of territorial marketing, and fight the overloading of natural and cultural attractions [6].

To highlight the great potential of slow adventure, Varley and Semple [4] claim that "(...) *tourists of the twenty-first century seek unusual new luxuries in the form of time in nature, birchwood fires, cooking their own wild food, carrying their own luggage over rough lands or along remote coastlines in kayaks*". Also, Varley, Farkic and Carnicelli [5] agree that "(...) *stating that slow adventure journeys include elements such as cooking wild food, gathering around a campfire, telling stories, being relaxed and comfortable with where you are and with who you are*". Such experiences are prized and carry a high price tag in the marketplace, as they are currently a scarce resource of rich, meaningful, potentially transcendent and intense experiences [4].

Besides, the authors mentioned above highlight that many adventure clients take 'time out' and value this as part of their peak experience, beyond the easy-to-retell stories of risk, danger, and 'extreme' activity [4, 5]. Nevertheless, the literature also tell us that these experiences can only adequately be delivered by well-trained, professional guides who, in addition to the hard skills of navigation, first aid, mountaineering, or kayaking, must be versed in the soft skills of outdoor hospitality, emotional intelligence, and facilitation. That means that the skills required to deliver high-quality slow adventure experiences are considerable. However, they are also valued by many sectors of late-modern society and are regarded as potentially lucrative [4, 5].

All in all, slow adventure is a new marketing concept that builds the experience around the participants' strong commitment and their interaction with nature and transformative moments. Correctly exploring the concept of slow adventure can be of great importance to remote rural nature-based destinations. However, it is essential to develop skills and train professionals properly [7].

2. The Azores archipelago in a nutshell

The Portuguese autonomous region of Azores is one of the European Union's Outermost Regions (OR). It includes nine islands positioned in the middle of the Atlantic Ocean, a midpoint between the North American Continent and Europe (**Figure 3**). Thus, this insular region is strategically placed, and it has

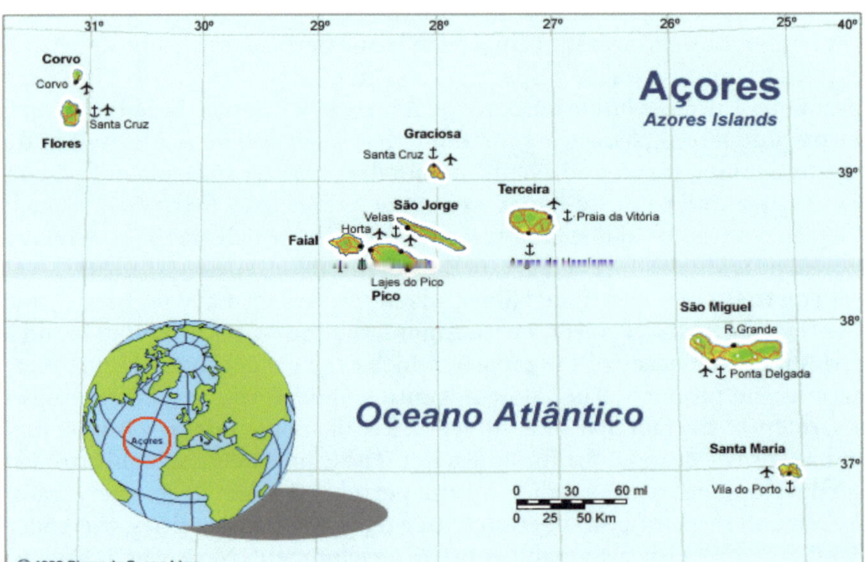

Figure 3.
Azores archipelago location [8].

been recognized as a sustainable nature-based destination. Newly, based on its remoteness and breathtaking nature, it has been frequently classified with adventure tourism by worldwide references like Bloomberg, Departures, BBC, Forbes, GeekyExplorer, Lonely Planet, are just a few examples. Also, its natural and cultural heritage and its generalized rural environment make it the perfect place for slow adventures [7, 9–11].

The islands have exceptional weather and soil conditions that have directly influenced the socio-economic development of the region. Their volcanic origin has resulted in a peculiar and rugged terrain and a stunning geological diversity complete with beautiful landscapes. In the Azores, there are 26 active volcanic systems; eight of them are underwater [12]. Adding to this, three tectonic plates (North American, Eurasian and African) meet in the Azores territory, where there is also a geological rift (Terceira's Rift). All these combined results in intense seismic activity and very diverse geological occurrences. The region's environmental and landscape richness, characterized by the intense green color, is also influenced by a maritime subtropical climate, with mild temperatures all year long (and a low-temperature range), steady rain, and significant volatility of the weather conditions [7, 12].

The nine islands combined have more than 2.000 km2. The islands are distributed in a 600 km axis (southeast-northwest direction), granting the region one of the bigger Exclusive Economic Zones (EEZ) in the EU [7, 13]. The distance to Lisbon, the Portuguese capital, is around 1.400 km (two-hour flight), and the distance to Boston (USA) is about 3.300 km (five-hour flight). For this reason, the Azores has been considered as an OR of the EU, alongside Madeira (Portugal), the Canary Islands (Spain), Guadeloupe, French Guiana, Réunion, Martinique, Mayotte, and Saint-Martin (France), under Article 349 of the Treaty on the Functioning of the European Union [8, 14].

Some of the Azores' main challenges arise from this distance to the main decision centers, beyond its internal market fragmentation and geographic dispersion. The archipelago has natural constraints that demand constant attention, similar to what occurs with another Portuguese Insular Region, Madeira Island [15–20]. The heterogeneous territory's fragmentation resulted in very different islands in terms of area and natural resources, with significant land dispersion. For these reasons, five islands—Santa Maria, Graciosa, São Jorge, Flores, and Corvo—have been categorized as the "Cohesion Islands" (being the smallest islands or more significant challenges to their development) and benefit from positive discrimination in the regional economic policies. The remaining four islands—São Miguel, Terceira, Pico, and Faial—are the most developed, although substantial differences exist among them. This context has apparent impacts on the regional economy, including effects on resource efficiency, population concentration, internal market dynamics, and the need to have multiple structural infrastructures, such as ports, airports, health units, or other public services [21].

In 2017, the GDP of the Azores GDP 4.067 million euros (at market prices), making it one of the poorest regions of the European Union (<75% of the average GDP per capita UE-27) [21]. Public administration has a considerable impact on economics and concentrates a large portion of the jobs. Agriculture and fisheries are historic structural activities in the Azores, and some of their produce, including milk, are vital for local industries, such as dairy. Additionally, tourism is one of the more important economic activities in the region and has been gaining ground in the last few years. Notwithstanding, people in the Azores have low education levels, and unemployment has been high and very difficult to overcome in recent years [21].

Despite the numerous challenges listed above, the islands' geographical location has also granted regional development exciting opportunities. Exceeding a new potential for space research and exploration, historically, many transatlantic routes have passed through the Azores, making the islands critical in logistics support to sea and air navigation, both in military and commercial activities. Eventually, it should also be emphasized that the big size (almost 1 million km2) of the EEZ grants the Azores unique opportunities regarding the Blue Economic, logistics, fisheries, nautical tourism, and scientific research and exploration of marine resources [21].

Tourism is perceived as a sector of high strategic importance for the Azores, which can contribute decisively to the development and growth of the region [22]. The natural and cultural resources of the Azores are varied and unique. They are the base for incredible adventures and extraordinary experiences, many of which are fast-pace high adrenalin activities. In contrast, others are very relaxing and dig deep into the social and cultural roots of the local people.

Besides, in 2015, it was a paradigmatic shift in tourism in the Archipelago. The air transportation paradigm was partly liberalized, stopping the local public airline's long-time monopoly and conceding the origin of the operation of low-cost airfare groups (Ryanair and Easy Jet). Therefore, the amount of tourists has been growing, promoting a notable novelty in the local market [23]. Hence, the local economy is developing, and investment is being used in tourism activities, varying from accommodation to entertainment.

Concerning accommodation, hotels and related businesses have a large majority of the bed supply and contemplate approximately 78% of the demand. So, looking at this specific accommodation type's evolution, it becomes easy to understand the Azores tourism sector's overall growth. Although, it should also be noticed that other kinds of accommodation (rural tourism, self-catering houses, hostels, or guesthouses) are increasingly more crucial to the genuine experience of the Azores (**Figures 4** and **5**) [21, 22].

Notwithstanding all the innovations that have been running on, there are still fundamental obstacles to overcome, as is the seasonality's case very pronounced, as it can be seen through the number of guests in hotels and related businesses. Visitors are endeavoring the Azores, particularly in the summertime, when there weather are more stable. Thereby, strategic planning needs to be meticulously studied, and sustainability problems are pivotal in this task [7]. Due to the necessity to preserve the Azores' competitive benefits, it is imperative to build an adequate model and

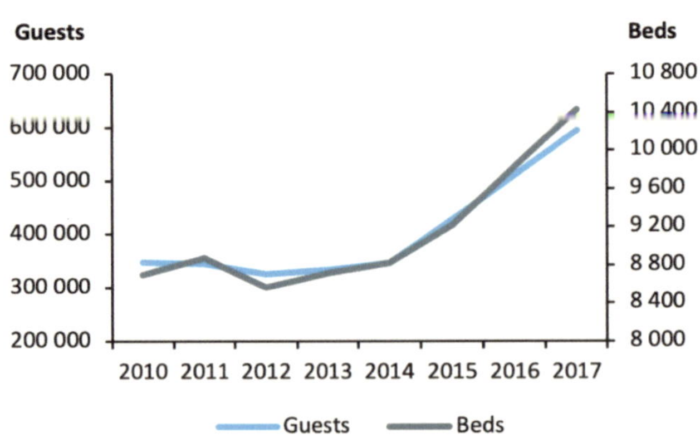

Figure 4.
Graphic of the evolution in the number of guests and in the number of beds in hotels and related businesses [23–30].

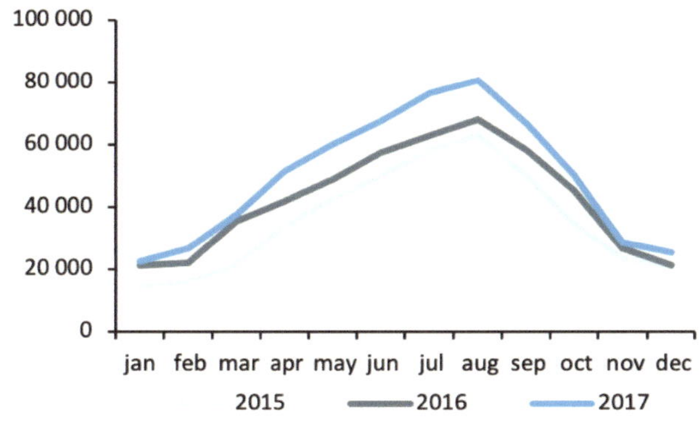

Figure 5.
Guests in hotels and related businesses [27–29]. X – Refers to the time frame (months); Y – Number of guests.

keep developing new products that grant more value increasingly to the Azores' experiences as a way to place the destination, respectively. In this regard, nature, slow, and adventure tourism can be frontline outputs of this model.

3. The Azores tourism strategy

The development strategy for Azores tourism is outlined in the PEMTA (Strategic and Marketing Plan of the Azores Tourism) [31], openly unveiled in 2016 by the Regional Government. The document points the fundamental intentions to be accomplished, the destination's market positioning policies, and the tourism outcomes that should maintain this strategy.

Thus, PEMTA fixes four decisive aims for Azores tourism. Such goals are in order amidst the life sequence stage of the destination, its resources, tourism agents, and supply and market dynamics: i) Support the Azores knowledge to the final consumers; ii) Place the Azores as an exclusive destination of enthusiastic nature; iii) Encourage continuous cooperation among public and private actors on the implementation of the plan; and iv) Increase the destination's competitiveness and boost tourist flows [6].

Moreover, five key principles were identified to accomplish these purposes. Those are the constructs of the strategy, managing all the decisions and actions to be disposed of: (i) nature tourism is the principal tourism product of the Azores, profiting on its natural resources and biodiversity, although it requires an equivalent products strategy; (ii) guarantee the feasibility of visiting all the islands, getting advantage of the reforms in the inter-island transport system; (iii) fix as distinguishing characteristics the closeness to the market, the singularity and the authenticity of every island, the landscape, the safety and security and the peace of the locals; (iv) promote the constant development of the performance of the tourism sector's particular areas; (v) act on the Azores sustainability, protecting the places and the local communities, by encouraging the principles of sustainability [6].

In fact, PEMTA proves that the Azores' market positioning should be based on the idea of a natural destination and unique excellence without outside influences [31]. Also, it should not have a mission to mass tourism; however, it should rather be centered on precise visitor niches that endeavor once in a lifetime experiences. Therefore, the following eight points should be comprehended: i) A Portuguese (European) destination in the heart of the Atlantic Ocean; ii) Environmentally protected volcanic islands of prolific nature; iii) Harmony of the four elements: water, earth, fire, and air; iv) Exclusive; v) Mystic beauty; vi) The visitor is welcome as a distinctive guest; vii) Safety and security; and viii) Diversity and quality of the sea and land ventures [32].

Furthermore, the Regional Government announced the POTRAA (Spatial Plan of Tourism of the Azores' Autonomous Region) in 2008. It was established as the primary tool to reach the sustainable development of the region's tourism sector. Consequently, it was also an instrument to lead the diverse economic actors and control the administrative action, setting the strategic tourism products and the tourism supply progression until 2015 [6].

POTRAA enabled the combination of much more than spatial management tools, expanding its influence to all the region's tourism strategy. Recognizing that the perception of Azores as a destination was heavily linked to nature and sustainability, the plan set six Strategic Development Guidelines in order with the SDG (Sustainable Development Goals), which were: SDG 1 – Reinforcement and developing of technical, organizational, and regulatory requirements to support tourism development; SDG 2 – Support to the modernization of infrastructure and complementary/support services of the tourism sector, to the internal and external accessibilities and the tourism signage; SDG 3 – Support to the development, qualification, and diversification of the regional tourism supply; SDG 4 – Incentive the tourist demand and the external awareness of the regional tourism; SDG 5 – Support for specific actions respecting the spatial planning of tourism; SDG 6 – Support to the implementation, follow-up, and evaluation of the plan [6].

POTRAA set a Territorial Organization Model, supported by cartography, surpassing proposing a prototype to administer the tourism supply in each island and a coordinated approach with municipal territorial management tools. The plan also recognized each island's principal tourism products and strategically ascertained a genuine core tourism product and some other corresponding characteristics. Such products and resources should lead the regional tourism strategy and the local tourism growth according to the islands' singularities and particular social, economic, environmental, and cultural circumstances [6].

According to Silva and Almeida [32]: "(...) *regions such as the Azores have great potential for the development of nature and adventure tourism products, but, due to their early stage of development as destinations and the limitations in their carrying capacity, it is necessary to guarantee a sustainable tourism development model*". This model should be directed on the endogenous resources to reach the diversification of tourism products, convince visitors' expectations, and minimize the impact of tourism activities. Hence, a solid destination planning and tourism development management show to be from pivotal importance to avoid some of the impacts linked with the development and overload stages of a destination lifecycle, like contaminated environments and landscapes, the decline of heritage, water pollution, erosion, and traffic jam. Once, slow adventure matches excellently in this situation and can also be a vehicle to reach sustainable and sustained tourism growth.

In fact, sustainability in the Azores Archipelago is practiced very thoughtfully by the regional authorities, and it is admitted as a decisive factor for the region's tourism sector. Back in 2007, National Geographic Traveler named the Azores as the Second-Best Islands for Sustainable Tourism. The report included very relevant assessments of the natural environment and cultural variety. Still, it was also pinpointed that unsuitable development was starting to emerge.

Several awards have been granted to the Azores Region. Among them are international evaluations regarding sustainable development and sustainable tourism. Some of these prizes come from important international institutions, such as UNESCO, National Geographic Traveler, QualityCoast, and Green Destinations (**Table 1**). In 2018, the Azores has conferred the Best Sustainable Destination of the Atlantic award, a significant milestone for the regional tourism sector.

Award	Organization	Comments	Year
Top 10 Most Sustainable World Destinations	Green Destinations	Best of the Atlantic	2018
Top 100 Most Sustainable World Destinations	Green Destinations	First place in 2014, with 8.9 points out of 10	2018
			2017
			2016
			2014
QualityCoast Platinum Award	QualityCoast – Coastal and Marine Union of the European Union	2017: On par with two other Dutch destinations – Goedereede and Westvoorne	2017
		2014–2016: First and, at the time, only destination with this award	2014–2016
QualityCoast Gold Award	QualityCoast – Coastal and Marine Union of the European Union	Best Quality Coastal Destination in Europe	2013
Best of the Best – Nature Award	European Commission	Granted to Project "Life Priolo," which was developed between 2003 and 2008, focused on the protection and restoring of the risk vegetation of the laurel forest of the Azores	2010
Second Best Islands in the World for Sustainable Tourism	National Geographic Traveler	—	2010
Ospar Convention	OSPAR Commission	12 locations identified for the protection of the maritime environment of the Northeast Atlantic	2010
Biosphere Reserves	UNESCO	Island of Flores	2007
		Island of Graciosa	2007
		Island of Corvo	2009
Natura Network 2000	European Commission	23 Special Preservation Areas	1989
		15 Special Protection Areas	
		2 Important Locations of the Community	
UNESCO World Heritage	UNESCO	Landscape of Pico Island	1983
		Vineyard Culture	2004
		Historical Centre of Angra do Heroísmo	

Table 1.
Azores sustainability and sustainable tourism awards [20].

4. The tourists profile and the international attention for the region

Azorean officials have been carrying out essential attempts to accurately comprehend the demands and identify the tourists' profile that visits the Archipelago. Additionally, to work done during PEMTA's [30] progress, there are currently two principal actions that make possible the monitoring of the tourists' profile and behavior. Thus, the Azores Tourism Observatory (OTA) uses surveys to evaluate the satisfaction of tourists tourists'the Region. Also, the Regional Government, in synergy with the GMT Hospitality, has been an online reputation monitorization study since 2016, enabling the perception of tourists' behavior in the Archipelago. These two tools merged present a compelling basis for the correct understating of the demand and the precise contingencies to classify new tourism growing opportunities.If we focus on OTA's 2017 report [33], it shows that tourists are fascinated with the Azores as well as with the quality of the experience in the destination surpasses their original expectations. It also points pertinent data about the tourists' profile, consumption behavior, and impulses.

In this regard, the medium age of tourists that visited the Archipelago in 2017 was 42. The preponderance was male (53,2%) and married (52,3%) with high qualifications, once 42% claim to possess a master's degree or a Ph.D. Several of them were on medium or leading professional positions (44,1%) and had an average monthly income of around 1.500€. In fact, the information about the income has a powerful influence from Portuguese tourists. These tourists comprise approximately 50% of the demand and have an average monthly income of approximately 900€ - much lower than, for example, German tourists (more than 1.800€) [33]. Furthermore, the report also shows that tourists commonly visit the Azores as their primary option (82,3%) for holidays and leisure (85,8%), and numerous of them are first-timers (83,9%). However, they expect to return (63,6%). Normally, the purpose of visiting the Azores evolved from word of mouth endorsement from colleagues, friends, and family (36,5%), yet the internet also played a pivotal role (35,6%). These tourists habitually planned their travel from one to three months before (47,6%), and they traveled with their spouse or partner and their children (70,3%) [6, 33].

The enthusiastic natural resources and the essence of the Azores' experience have become very appealing factors for international travelers and the tourism press. In the last few years, several reports about the Azores as a tourism destination and its potential for adventure have been published. Associations with iconic destinations, such as Iceland, Hawaii, and New Zealand, have been regularly made among many other paradisiac locations and have contributed to the region's expanding awareness [21]. In the beginning, the European market was the central objective, although lately, the North American market is considering what is going on in the Azores Region. According to Couto et al. [21]: "*Sustainability, nature, food, and local culture are essential factors in this context and grant the Azores an almost idyllic image for great holidays and travels.*" Nonetheless, adventure experiences and the diversified tourism activities are also earning much publicity, addressing the Azores as one of the most desirable locations to experience some of them, including whale watching, canyoning, or hiking and diving. Beyond sustainability and sustainable tourism prizes and certifications, the Archipelago has been awarded other grants concerning tourism and its endogenous resources. Some of these awards are also very impressive when reflecting on the Azores Region's new adventure tourism experiences and products [21].

5. Closing section

Based on the above section, it is possible to understand that the Azores Region's slow adventure is a perfect match. The Azores Islands are peaceful, secure, and comprehensive of nature to explore. The Archipelago's nine islands provide great diversity to the destination, and the tranquil rural landscape fits the ideal framework for unparalleled, charming, and relaxing experiences. The islands' volcanic nature makes them perfect places to explore, with stunning landscapes, unique occurrences, unexpected phenomena, and incredible examples of human creativity and perseverance. An exciting game seems to continuously unfold, as the green of the land and the sea's blue guide the local people's everyday lives [21].

Similar to other studies [7, 9, 10, 13], the Azores tourism strategy embraces nature as an essential resource for its development. In this sense, culture is also a priceless asset [7, 21]. Thereby, these two assets connected give different possibilities for adventure tourism products and a long-term strategy structure. Sustainability has also been viewed as a decisive factor for Archipelago development and has an indispensable role in the regional policy [7–12, 14–21, 34].

Another critical factor, previously verified in other peripheral territories, as borderlands, is the marketing and advertisement [35–37]. This factor in the region has been placing the Azores as an adventure destination [21]. According to Couto et al. [21]: "*The conventional adventure has been the focus, but the international press has also given attention to products that fit the slow adventure concept, like beekeeping, slow volcanic cooking, weird seafood, wellness and hot springs, wild nature experiences, family travel, hot water baths in the sea, good wine, hiking in volcano craters, experiences in fishing boats and agritourism.*"

Moreover, tourists that visit the Azores Archipelago have character traits that allow slow adventure an unprecedented potential [21]. Meanwhile, companies are expanding their portfolios, although they are still needing in delivery and marketing ability and can also benefit from even more concepts for new products. Nevertheless, these are great possibilities to profit on the SAINT project's methodology, intending to promote local companies and match the demand's expectations [21, 33].

In short, it is secure to state that the Azores Archipelago has a huge potential to welcome the slow adventure idea and promote products and experiences beneath the international label driven in the SAINT Project [21].

Funding

This research is financed by Portuguese national funds through FCT – Fundação para a Ciência e a Tecnologia, I.P., project number UIDB/ECO/00685/2020 and also by the project GREAT - Genuine Rural Experiences in the Azores Tourism with the code: ACORES-01-0145-FEDER-000089.

Author details

Pedro Pimentel[1*], André Oliveira[2], Gualter Couto[1], João Crispim Ponte[2] and Rui Castanho[3,4,5,6]

1 School of Business and Economics, University of the Azores and CEEAplA, Ponta Delgada, Portugal

2 Fundo de Maneio, Ponta Delgada, Portugal

3 Faculty of Applied Sciences, WSB University in Dąbrowa Górnicza, Poland

4 VALORIZA - Research Centre for Endogenous Resource Valorization, Portalegre, Portugal

5 CITUR - Madeira - Centre for Tourism Research, Development and Innovation, Funchal-Madeira, Portugal

6 CNPQ Research Group Aquageo Ambiente Legal, University of Campinas (UNICAMP), Campinas, Brazil

*Address all correspondence to: pedro.ms.pimentel@uac.pt

References

[1] *AM Reports, Volume Nine – Global Report on Adventure Tourism*, UNWTO, Madrid, UNWTO (2014)available at https://bit.ly/2HxnY3z.

[2] Buckley R. Adventure tourism research: A guide to the literature. Tourism Recreation Research. 2006;**31**(2):75-83

[3] Buckley R. Adventure tourism products: Price, duration, size, skill, remoteness. Tourism Management. 2007;**28**(6):1428-1433

[4] Varley P, Semple T. Nordic slow adventure: Explorations in time and nature. Scandinavian Journal of Hospitality and Tourism. 2015;**15**(1-2):73-90

[5] Varley P, Farkic J, Carnicelli S. Hospitality in wild places. Hospitality & Society. 2018;**8**(2)

[6] Couto, G., Pimentel, P., Ponte, J., Oliveira, A., Gonçalves, P. Arruda, A., and Oliveira, B. (2020). Adventure Tourism in the Azores. Ed. Centro de Estudos de Economia Aplicada do Atlântico (CEEplA). ISBN: 978-989-8870-27-8

[7] Castanho, R.A., Couto, G., and Pimentel (2020). Principles of Sustainable Tourism and Cultural Management in Rural and Ultra-peripheral Territories: Extracting Guidelines for its application in the Azores Archipelago. Cultural Management: Science and Education (CMSE) ISSN: 2512-6962 DOI: 10.30819/cmse.4-1.01

[8] Castanho RA, Behradfar A, Vulevic A, Naranjo Gómez J. Analyzing transportation sustainability in the Canary Islands archipelago. Infrastructures. 2020;**2020, 5**:58. DOI: 10.3390/infrastructures5070058

[9] Ponte J, Couto G, Pimentel P, Oliveira A, De Maneio PDF. Tourism activities and companies in a sustainable adventure tourism destination: The Azores. Tour. Manag. Stud. 2018;**14**:25-38

[10] Couto G, Castanho RA, Sousa A, Pimentel P, Santos C, Carvalho C. The impacts of COVID-19 crisis over the tourism expectations of the Azores archipelago residents. Sustainability. 2020;**2020, 12**:7612. DOI: 10.3390/su12187612

[11] Castanho RA, Couto G, Pimentel P, Carvalho C, Sousa A. Territorial management and governance, regional public policies and their relationship with tourism. A case study of the Azores autonomous region. Sustainability. 2020;**2020, 12**:6059. DOI: 10.3390/su12156059

[12] Azores Green Mark (2020) Formation and Main Eruptions. Available at: https://www.azoresgreenmark.com/acores/formation/ (Acceded at September 17, 2020).

[13] SAINT Project (2019). SAINT Project. Available at: https://saintproject.eu/. (Acceded at December 15, 2019)

[14] Naranjo Gómez JM, Lousada S, Garrido Velarde J, Castanho RA, Loures L. Land use changes in the canary archipelago using the CORINE data: A retrospective analysis. Land. 2020;**2020**(9):232. DOI: 10.3390/land9070232

[15] Lousada, S., and Castanho, R.A. (2020). Cooperation Strategies Towards Sustainability in Insular Territories: A Comparison Study Between Porto Santo Island, Madeira Archipelago, Portugal and El Hierro Island, Canary Archipelago, Spain. Chapter in the book:

Strategies for Sustainable Development. Editor of the Book. IGI GLOBAL Pp 216-238. ISBN13: 9781799825135. DOI: 10.4018/978-1-7998-2513-5.ch012

[16] Lousada S, Silva P, Castanho RA, Naranjo Gómez J. Modelação de sistemas de abastecimento de água. O caso de Ilha da MadeIra. Modelling water system supplies. The case of Madeira island. Bitacora Urbano Territorial. 2019;**29**(2):89-98. DOI: 10.15446/bitacora.v29n2.69381

[17] Castanho, R.A., Lousada, S., Naranjo Gómez, J.M., Escórcio, P., Cabezas, J. Fernández-Pozo, L. and Loures, L. (2018). Dynamics of the Land Use Changes and the Associated Barriers and Opportunities for Sustainable Development on Peripheral and Insular Territories: The Madeira Island (Portugal) [Working Title], IntechOpen, DOI: 10.5772/intechopen.80827. Available from: https://www.intechopen.com/online-first/dynamics-of-the-land-use-changes-and-the-associated-barriers-and-opportunities-for-sustainable-devel

[18] Lousada S, Escórcio P, Castanho R, Loures L, Naranjo Gómez JM, Fernández-Pozo L, et al. A multi-variated analysis on ultra-peripheral territories. A case study approach: Sustainability indicators proposal for Madeira Island. Chapter in the book: PLANEAMIENTO SECTORIAL: RECURSOS HÍDRICOS, ESPACIO RURAL Y FRONTERAS. Ed. Thomson Reuteurs Aranzadi. Pp. In: 293-303 ISBN 978-84-1309-065-8. 2018

[19] Castanho, R.A., Lousada, S, and Camacho, R. (2018). Modelling Torrential Rain Flows in Urban Territories: Floods. The Case Study of Madeira Island. 25th APDR CONGRESS. FCHS, Lisbon-Portugal. July 5-6. Circular Economy, Urban Metabolism and Regional Development: Challenges for a Sustainable Future. Faculty of Social Sciences and Humans (FCSH) da University NOVA – Lisbon. Proceedings Book. ISBN 978-989-8780-06-5. pp. 263-267. pp. 378-388.

[20] Gonçalves R, Camacho R, Lousada S, Castanho R. Modeling of maritime agitation for the design of maritime infrastructures: The case study of madeira archipelago. Revista Brasileira de Planejamento e Desenvolvimento, Curitiba. 2018;**7**(1):29-50. DOI: 10.3895/rbpd.v7n1.7136

[21] Couto G, Pimentel P, Ponte J. Tourism development potential in an insular territory: The case of Ribeira Grande in the Azores. Journal of Tourism Research & Hospitality. 2017;**6**(2)

[22] Vieira J, Couto G, Pimentel P, Menezes A, Moniz A, Sousa F. The satisfaction of the Nordic tourist with the Azores as a destination. Scandinavian Journal of Hospitality and Tourism. 2014;**13**(1):58-72

[23] SREA. Estatísticas do Turismo janeiro a dezembro de 2010. Açores: Serviço Regional de Estatística dos Açores; 2011 available at https://bit.ly/2HvmeYD

[24] SREA. Estatísticas do Turismo Janeiro a Dezembro de 2011. Açores: Serviço Regional de Estatística dos Açores; 2012 available at https://bit.ly/2HvmeYD

[25] SREA. Estatísticas do Turismo janeiro a dezembro de 2012. Açores: Serviço Regional de Estatística dos Açores; 2013 available at https://bit.ly/2HvmeYD

[26] SREA. Estatísticas do Turismo janeiro a dezembro de 2013. Açores: Serviço Regional de Estatística dos Açores; 2014 available at https://bit.ly/2HvmeYD

[27] SREA. Estatísticas do Turismo janeiro a dezembro de 2014. Açores: Serviço Regional de Estatística dos Açores; 2015 available at https://bit.ly/2HvmeYD

[28] SREA. Estatísticas do Turismo Janeiro a Dezembro de 2015. Açores: Serviço Regional de Estatística dos Açores; 2016 available at https://bit.ly/2HvmeYD

[29] SREA. Estatísticas do Turismo janeiro a dezembro de 2016. Açores: Serviço Regional de Estatística dos Açores; 2017 available at https://bit.ly/2HvmeYD

[30] SREA. Estatísticas do Turismo janeiro a dezembro de 2017. Açores: Serviço Regional de Estatística dos Açores; 2018 available at https://bit.ly/2HvmeYD

[31] Instituto de Planeamento e Desenvolvimento do Turismo IPDT–. Plano Estratégico e Marketing do Turismo dos Açores. Governo Regional dos Açores. In: available at. 2016 https://bit.ly/2vs1hZJ

[32] Silva F, Almeida M. Sustentabilidade do turismo na natureza nos Açores – o Caso do canyoning [sustainability of nature tourism in the Azores – The case of canyoning]. In: Almeida M d C, editor. Turismo e desporto na natureza [Tourism and sport in nature]. Estoril, Portugal: Associação de Desportos de Aventura Desnível; 2013. pp. 5-19

[33] OTA – O d T d A. Satisfação do Turista que Visita os Açores – Ano 2017. Observatório do Turismo dos Açores. In: available at. 2018 https://bit.ly/2vxcsjW

[34] Sousa Lima, P. (1999). Mapa dos Açores. Available at: http://diveazores.blogspot.pt/2006/01/mapa-dos-aores-azores-map.html (Acceded at September 17, 2020).

[35] Castanho RA, Vulevic A, Naranjo Gómez J, Cabezas J, Fernández-Pozo L, Loures L, et al. Assessing the impact of marketing and advertising as strategic approaches to Eurocities development: An Iberian case study approach. European Journal of International Management. 2019 ISSN: 1751-6765

[36] Vulevic A, Djordjevic D, Castanho RA, Cabezas-Fernandez J. Social marketing and their related challenges for the limited access for people living with a disability: A Serbian case study. Chapter in book: Case studies on social marketing. Ed. springer. In: 10.1007/978-3-030-04843_19. 2019

[37] Vulevic, A., Castanho, R.A., Gómez-Naranjo, J., Cabezas, J. and Fernández-Pozo, L. (2017). Transport Accessibility to Achieve Sustainable Development. Principles for its Application in CBC Projects: The Euro-City Elvas-Badajoz. Proceedings Book - 16th Congress on Public and Nonprofit Marketing (IAPNM). Conference Proceedings First Ed. 2017. Badajoz. ISBN: 978-84-697-5643-0.

Section 4

Rural Tourism and Regional Development

Chapter 6

Rural Development and Rural Tourism: The Impact of Infrastructure Investments

António Almeida and Luiz Pinto Machado

Abstract

Rural and peripheral development is still a matter of concern in several western countries. Depopulation, low density of business activities, younger people emigration and better-qualified individuals feeling that such regions have been abandoned by the government, and incapable of moving on, are among the key indicators to "understand" rural and peripheral areas. Rural tourism has long been understood as an effective catalyst of change in depressed and deprived (of entrepreneurial capacity) areas and to explore a unique set of amenities. Because of funds directed to help private investment projects in rural tourism facilities, most peripheral areas are now relatively well endowed with key infrastructures. Nevertheless, the tourism lead approach produced mixed results due to low levels of demand in some areas and lack of a cooperative behavior among providers to maximize the opportunities offered by the wide range of attractions. In this paper, we investigate to what extent investments in infrastructure helped the rural tourism sector to attract more visitors in Madeira. Based on the panel-data approach, this paper provides insights to analyze the development path of rural tourism in Madeira and to explores how local policy makers may be the "missing link" needed to improve the sector prospects based on tangible and intangible amenities.

Keywords: rural tourism, panel-data, investment in infrastructure, nature based tourism, regional development

1. Introduction

The economic development of rural areas is still a pressing issue in most western countries, especially in peripheral areas coping with high levels of unemployment, declining farm income, depopulation and emigration of the younger and better-qualified individuals [1, 2]. In general, such regions suffer from limited options with regards to economic development outside agriculture [3–6], which led policy makers to re-think strategically the social and economic fabric of rural areas [7]. Policy measures aiming at diversifying rural economies through the development of new sectors/products [2, 8] have also been conceptualized by regional governments on islands, in an attempt to reverse the trend of decline in terms of quality of life that has been felt across rural areas. Such approach is strongly encouraged by the European Union, to "fill the gap" between the over-developed coastal areas and the rural hinterland [9].

To improve the competitiveness of the primary sector and protecting the environment and countryside, several measures were introduced as the third axis of the EU rural development program that aimed to enhance the quality of life in rural areas and to encourage the diversification of economic activity [10], in order to improve the rural area's capacity to provide goods and services demanded by the wider society, and to invest in the livelihoods of those residents in rural areas. According to the European Evaluation Network for Rural Development (ENRD), axis 3 consists of three modules aimed at (i) sustainable eco-nomic development, which includes the diversification of farm holdings and rural tourism promotion; (ii) life quality and viability of the rural community; and (iii) the acquisition of specific skills [11–13].

Past research suggests that agro-tourism and rural tourism RT have played an important role in the development and revitalization of the economic base of deprived and marginal areas partially based on "positive linkages" between agriculture and tourism [10, 14–16].

RT enables a complete tourism experience with representation of accommodation and attractions. On the other hand, creates spatial attachment on visitors and encourages them to some loyalty. RT helps to product enlargement and innovation and can draw in new capital form for a region and countries [17].

Today, RT is supported in all EU countries due to its socio-cultural, economic, spatial and environmental functions and positive effect on developing vernacular building sector in rural area [18].

There is also evidence to suggest socio-cultural gains in terms of villagers "adding value to families' capabilities and skills, enriching their lives in non-pecuniary ways" [1, 10, 19–21], but most studies tend to be the result of analyses applied to the context of mainland areas in Western Europe. Zasada and Piorr [10] analized the German Bradenburg area and concluded that generally, diversification into non-farming activities represents, to some extent, a rarity in the rural periphery of Brandenburg, where large agri-businesses and co-operatives prevail, and where the demand for rural goods and services is limited. They concluded, in contrast, that measures where public authorities play a more important role, such as in tourism development and village renewal, show a stronger responsiveness to framework conditions, especially to the rural community characteristics. This indicates the effect of substantial political targeting. As the village renewal measure is absent in urban and economically advantageous communities, it truly represents a rural domain for addressing disparities among regions with structural weaknesses and are prone to demographic change.

In a Different vision Lekakis and Dragouni [22] explore the texture and values of rural heritage, by drawing on empirical evidence from Naxos island in Greece and critically examines the character and significance of this composite past that is made up from an assemblage of tangible and intangible elements interwoven with the island's agricultural life as recently as the mid-20th century, the man-made and natural components of the islands' rural landscape form together a heritage 'in the making', a process of bottom-up heritage designation by the surrounding communities, an act of 'mnemeiosis' outside the official narrative and cultural management practice. As Madeira Island offers an interesting and unique Natural heritage (part is UNESCO heritage), reinforced by the original rural landscape characterized by small plots of cultivated land, well demarcated and arranged in terraces that unfold to sea level, can add some value to the countryside attractiveness.

López-i-Gelats et al. [23] assets that farm diversification have become prevalent throughout the European countryside", and Fleischer and Pizam [21] considers RT as "the dominant factor in the rural economy of [some areas in] Western Countries". According to Briednham and Wichens [24] "RT is increasingly viewed as a panacea, increasing the economic viability of marginalized areas, stimulating social regeneration and improving the living conditions of rural communities" [24]. However, with few exceptions, there has been very little research about tourism development in the rural hinterland on islands.

There is an obvious bias in writing on RT towards the Western European, which can be easily explained by the long established tradition of farm tourism in the United Kingdom, Germany, France and Austria [23]. Consequently, the possibility of transferring the theoretical models of RT development is problematic, "at least not without substantial modifications". In fact, there are only a few empirical studies testing the tourism led growth hypothesis for rural areas on islands. Nonetheless, the normative discourse concerning rural areas on islands is impregnated.

Further research is also needed to analyze the impact of the absence of a central government policy on direct funding of RT facilities. By definition, RT is "specialist and small scale" [14] but also over-dependent on government subsidies. However, in times of declining public budgets, issues of financial sustainability are of the utmost importance and further investments must be linked to profitability rations and personal savings.

The main contributions of this paper, first is to analyze how the visitors′ attitudes towards farming is of the utmost importance to understand the alleged "shared destiny" of farming and RT in rural areas. Secondly, this paper elicits consumers′ preferences towards key attributes of the RT product based on a choice experiment, which has rarely been done before. We are especially interested in providing a measure of visitors′ preferences for farming activities, in order to provide useful insights to help to define priorities in terms of investment. Most papers assume a close linkage between agricultural activities and RT and explore the causal relationship running from farming activities to tourism development. In this paper, we explore RT development from an alternative point of view, as we investigate if rural tourism development can be accelerated by investments in farming facilities, or other types of infrastructure and complementary activities.

Both negative perceptions about the current economic status, due to Covid19, and positive expectations for the near future of rural areas in the service economy, knowing that tourists will look for more quiet, less frequented locations, preferably for non-shared use lead to an over-optimism about the impact of RT, with over-optimistic assumptions about the alleged potential of tourism to foster economic development.

This study offers an opportunity to document and understand the dynamics of RT development in the European periphery, a poorly studied geographical setting. Though some studies highlight the growing importance of RT on islands, the specificities of RT development on non-tourism spaces have not yet been rigorously examined in the literature.

2. Tourism, agriculture, and the regeneration of rural areas

It is now well documented that farming is not a key component of the RT experience, which means that the provision of a farming facilities may not be a necessary condition to offer a high-quality tourism experience. However, the sector needs to further meet visitors′ expectations and tackle the low occupancy rate problem. While it is doubtful that a "single silver bullet" (RT development) may spur economic dynamism in rural areas, small islands′ development options are limited, and clearly dependent upon agriculture and tourism. Consequently, there is an urgent need for experimentation in terms of local development approaches, business concepts and "alternative" attraction factors, that may or may not include offerings in terms farming facilities.

With regards to Madeira Island, RT emerged only in the 90s, with positive links between farming and RT yet to be developed. Tourism sector in Madeira reveals other interesting characteristics. Being one of the oldest tourist destinations in Europe, with about 30,000 Hotel beds distributed mainly on the south coast of the Island, it has surrendered to the dominance of the major tour operators, concentrating its tourist activity in its capital Funchal, and peripheral cities, forgetting for many years the countryside and the interior of the island.

In contrast to the mass tourism that seeks the great centralities, RT attracts minorities, especially individual tourists or small groups that seek tranquility and natural environment, staying in small hotel units or private houses (local accommodation, Airbnb etc..) and therefore a business uninteresting for large traditional tour operators.

Traditional RT, according to the terminology of Pina and Delfa [7] and Molera and Albaladejo [20], linked to homecoming connotations and city dwellers looking for short breaks [25], was never a key motivation attracting tourists to the region. Furthermore, rural tourism is not linked to farm tourism, as elsewhere in Europe [4, 26]. In most cases in Europe a rural house facility is whether a symbol of prestige and a subsidy-backed strategy to re-built the family heritage and property or just a normal business venture. Rural houses Owners cannot be considered as the "guardians of the countryside" [4], and while farmers in North-Europe may resist to the diversification due to an "anticipated loss of identity or social/cultural rewards" [4, 24], problems of "identity" on islands may arise because generally, managers have no experience in new information technologies and business models blending agriculture values with tourism services.

Regarding technology, in rural areas with limited infrastructure and technical knowledge, how could technology be made people more friendly to propagate its use needs to be discussed in depth [27], for example, how to have direct and easy connections with the new tourism distribution channels and allowing guests to book directly from their mobile apps to the owners, based upon accommodation offer, or through the new *apps* launched by strong web sites like Airbnb, VRBO, booking etc...,

As most islands economies only highlights in the tourism industry, attempts to diversify from the 3-S product are rather common nowadays [28–31]. However, an increase in the numbers of visitors and the development of new market niches does not depend only on bold and generic statements about innovative policies but above all on the effective enhancement of concrete policy measures. Quite often the "the lack of political mechanism to translate policy into practice" [2], and the "unusual context of small islands politics [32] along with the optimist and voluntaristic discourse about the alleged benefits of tourism are leading islands in the opposite direction, i.e., a "fatalistic path" in terms of over construction of infrastructures in coastal areas and overconsumption of scarce natural resources in rural areas compounded by returns below expectations in the new brand market niches [29].

In fact, several constraints have been identified in the literature. The first one regards the widespread adoption of a one size fits all approach, based on the Western European experience. RT in the United Kingdom, France, Germany and Austria evolved in association with farming-tourism based on the coexistence of farming activities and accommodation facilities. With regards to non-Western European geographical settings, recent studies provide mixed evidence regarding the level of symbiosis between tourism and agriculture and the extent to which RT may promote economic progress in depressed peripheral areas is still the subject of controversy [8]. Nonetheless, high levels of public funds have been poured in rural areas to support "the redevelopment of redundant farms buildings into accommodation facilities" [4] and to help to convert old family houses into accommodation facilities.

One of the important things to receive a vibrant and sustainable tourism in rural area is the conservation of rural cultural landscape and vernacular architecture to protect the natural and cultural values [33]. Consideration to the development as an issue on the agenda only could be expanded by the sustainability in natural and cultural benefits for their local communities. Developing this special tourism would be performed by responding to new markets, new lifestyles, and new product development opportunities [17].

Vernacular architectural structures in terms of contribution to the RT have been reviewed by [18], were this element seem to be capable to making an important contribution to RT, something that can be followed by the Madeira Island Authorities, as the typical architecture in Madeira is based in nice basalt stone houses with little touches of good taste in the doors and windows, with very comfortable interiors, some with nice fire places and terraces.

The voluntaristic approach adopted by most governments may not be completely illogical. Even if the tourism literature "is replete with seemingly contradictory observations" [34], tourism is an anchor for the islands' economy and, from a theoretical point of view, the development of a close relationship between agriculture and tourism seems rational, due to the size of the two sectors.

Since the adhesion of Portugal (and Madeira Island) to the EU in 1986, agriculture began to lose its importance as a source of employment and added value in the local economy. Agriculture today represents less than 2% of regional value added, but still about 10% of employment. The rural restructuring process has continued since then, and the Madeiran agriculture presents clear problems of competitiveness, linked the predominance of small-scale units and the difficult orography, being high protected from outside market forces and subsidized via EU programs.

Similarly, as Rodríguez and Pose [35] noted in Galicia, Madeiran agriculture is dominated by 'mini farm' model, that is, small patches of land divided from generation to generation. The most significant agriculture contribution undoubtedly concerns the ecological and social balance of the region. The extremely fragile nature of the ecosystem, along with an increased risk of the occurrence of natural disasters justifies the protection of agricultural areas.

In fact, the economic importance of agriculture lies within its subsistence value, crucial especially in times of Covid 19 crisis but also in the generation of aesthetical pleasant landscapes.

Another constraint faced by policy makers and entrepreneurs regards the specificity and originality of sector. Problems of lack of professionalism and skills gaps in critical areas of competence were also reported by Sharpley [14]. Farm diversification demands "new skills and competencies" and therefore a new "mentality and identity" Brandth and Haugen [24]. Another matter of concern regards the incompatibility of agricultural values with the provision of accommodation services. Most farmers exhibit difficulties in combining agricultural practices with guest-oriented values, which leads to the unwillingness or inability to apply a tourism led agenda. Burton also reports "loss of identity" associated with non-agriculture activities, and Sharpley [14], based on the analysis of RT development in Cyprus, highlights the difficulties to "combine the commodification of agricultural traditions through tourism with the industry of tourism", where there is no established tradition.

In general, the research available demonstrates that RT is a risky business [36], prone to "relatively poor financial success" [21] which is well exemplified by the over-reliance over financial assistance and subsidies [14, 25, 26, 37].

In Cyprus, Spain and Israel, financial assistance to help owners to restore and convert family houses and "unutilized farm buildings" into RT facilities was critical to 'persuade' managers to 'invest' in RT [14, 21, 38] found that "the financial returns in RT investments most often do not measure up to either the expectations of the politicians or that of farmers", and Barke [25] reports a high rate of failure at the early phases of the development of RT in Spain.

As "public assistance" is increasingly problematic, rural houses owners are required to finance further investments based on their own resources. Under such circumstances, decisions on investments tend to be based solely on expected returns and under-sized firms with limited financial capacity may feel constrained in making risky investments. In the opposite side, Zasada and Piorr [10] analyzed the German Brandenburg area and concluded that measures where public authorities play a more important role, such as in tourism development and village renewal, show a stronger responsiveness to framework conditions, especially to the rural community characteristics.

In the Small Economy - SME context, financial constraints favor investments with an immediate/visible financial return, and not distant from their "established technological" base. In fact, most firms operating in this sector, are "generally small, independent and family-run establishments", traditionally operated and "lacking the resources to promote themselves adequately and they have difficulties in adapting to current market mechanisms, which are becoming extremely competitive" [21, 39].

With regards to farming, most papers stress the increasing difficulties to apply a RT agenda informed by agricultural values. Some recent academic research suggests an ever-decreasing importance of agriculture to RT and other studies just see farming activity as a decorative and supporting role.

However, this does not mean the complete absence of a wide range of indirect links between agriculture and the quality of the tourism experience. Walford [40] suggests that agriculture indirectly relates to tourism via their contribution to esthetically attractive surroundings, and to the built-up of a relaxing atmosphere in the countryside and Frochot [41] asserts that "if farm and rural life are not consumed directly by all visitors, it probably remains a central component to the visual and social images of the countryside, particularly for urban dwellers". Farming produces valuable externalities of the "neighborhood effect", which amounts to an indirect effect of farming upon tourism. According to Fleisher and Tchetchik [42], artificial attractions and other firms operating in complementary sectors in rural areas "might be part of the experience and thus can benefit the entrepreneur".

On the contrary, visitors eager to learn and experience rural lifestyles would enjoy the liveliness of a typical village farm.

Osti and Cicero [43] revealed that tourists particularly enjoy the presence of a landscape comprising orchards, flowery/grassy meadows and vineyards.

Undoubtedly, RT can help make low-intensity agriculture more sustainable, while it can serve to attract new investors from the city or even foreigners interested in living in a calmer, more natural and safer environment. The question is to know, in addition to the rural landscape, the natural heritage and the typical local architecture (which must be protected with support that allows it to be enhanced), that other investments must be made in addition to farming. Improving the "levadas" system (channel irrigation network around the island with more than 2500 km) and mountain trails, making them safer and more accessible, can attract more tourists who love tracking's and mountain biking, in addition to bird watching, study of endemic flowers and plants among, in addition to the creation of viewpoints, sports and cultural spaces and support for the trade of typical food and drinks. Further research is also required to determine the ideal number of facilities that are required to guarantee a high quality experience. The industry´ actors would also be appreciative of further guidance on identify priority areas for investment.

3. Recent evolution of the rural tourism sector

The evolution of the number of establishments was multiplied by 15 between 1995 and 2019, which points to an annual growth rate of 12%. Based on the official statistics recorded since 1995, the RT sector has experienced considerable growth over the last 24 years, and the share of this market niche in the total number of visitors grown from 0,1% to 2,8%.

The traditional market for the region has been Western Europe, and in particularly, Portugal (mainland), Germany, United Kingdom and France and Scandinavia.

In the RT context, it's not only the number of visitors that matters, but also how these visitors are distributed over the year (**Table 1**). The period comprising June to September account for 39,1% of the total number of tourists concerning the RT sector and the same period corresponds to 42,7% of the total number of guests.

While the sector has succeeded in recording an impressive growth rate, occupancy levels are still quite low.

According to 2019 data, the RT sector had an occupancy rate of 39,8% compared to 58,0% for the sector a whole, which correspond to 68,56% of the former. The highest occupancy rate is achieved in August when half of the rooms were occupied. The average occupancy levels recorded in this market niche lead to a number of concerns notably in terms of the financial viability of the sector. While in several cases the sector can charge a higher price, data available suggests that the ADR (Average Daily Rate) for this market niches is 69,28€ compared with an average of 68,38€ for the tourism sector. Therefore, the prices charged by the sector are slightly below average, which appears in contradiction with the fact that the sector is aimed at the middle/high income market segment. While one may suggest that higher prices would lead to below average occupancy rate, this is not the case in Madeira.

	Total	Germany	France	Portugal
Jan	4,2%	4,5%	3,1%	4,4%
Feb	5,7%	6,8%	4,5%	6,0%
Mar	8,0%	10,3%	6,1%	5,9%
Apr	11,4%	11,6%	13,8%	8,7%
Mai	10,3%	9,8%	15,9%	6,0%
Jun	10,0%	10,1%	10,5%	10,5%
Jun	10,6%	8,7%	11,1%	11,8%
Aug	12,8%	9,1%	16,9%	15,5%
Sep	9,3%	9,4%	7,6%	9,7%
Oct	8,7%	10,9%	6,1%	9,2%
Nov	5,2%	5,5%	2,3%	7,0%
Dec	3,8%	3,3%	1,9%	5,4%
Jun to Sep	42,7%	37,3%	46,2%	47,6%

Source: Authors.

Table 1.
Four main markets in rural and seasonality tourism Madeira (2019).

Figure 1.
Length of stay in tourism and rural tourism in Madeira Island. Source: Authors.

The data available indicates that the RT sector has been able to create job opportunities in the rural hinterland. However, while the number of guests was multiplied by 67,5 between 1995 and 2019, the same figure for the number of jobs directly employed by the sector was multiplied only by a factor of 51,97. The data suggests that it takes an extra 1000 guests to get 5,3 extra jobs. Nevertheless, the same figure for the industry as a whole is 2,5, which suggests that the sector is more labour-intensive.

Data on length of stay indicates that on average each visitor stays for 3,72 nights (**Figure 1**). Visitors from abroad are highly likely to stay for a weeklong period, unlike Portuguese national that can stay for a couple days. Nevertheless, it is still difficult to explain such a low figure, unless we consider the visitors stays just for short period before moving out to another establishment elsewhere on the island.

Data on daily expenditure for the tourism sector suggests that visitors spend marginally on items such as cultural, sportive and entertainment relative activities. As the RT sector in intrinsically linked to a wide range of cultural, learning and outdoor it matters to check if rural visitors are more inclined to participate in such activities. Owners and operators may not know how to accurately evaluate benefits and costs of initiatives and measures aimed at increasing the overall quality of the product. In all evidence, they cannot charge higher prices to compensate for higher operational costs. The sector is under increasing pressure from the local lodgement sector. Data referring to 2019 shows that traditional hotel establishments (including hotels, apart-hotels, self-catering apartments, holiday villages and inns and Quintas da Madeira) account for 83,62%, the RT sector 3,31% and the local lodgement sector for 13,07% of the total number of beds.

4. Methodology

The data used in this study was obtained through a self-administered questionnaire survey carried out from August to November of 2018 in Madeira Island. To contact a large number of respondents, questionnaires were left with rural houses owners affiliated to the Madeira Rural Association (MRA).

More than 800 questionnaires were distributed across all the MRA members in the island and were collected and validated 360 questionnaires which constitutes an overall response rate of 45%.

Stata software and Discrete choice experiment model (DCE) was used to analyze the data, as this model is a technique for eliciting consumers´ preferences, provide valuable insights about the way individuals value the tourism product and the different attributes, that have been extensively applied to model and predict consumers´ behavior in different economic fields such as transport, economics, marketing, environmental and health economics. Louviere et al. [44] and Hensher et al. [45] provide a detailed analysis of the discrete choice methodology, which is based on the random utility theory and on the Lancaster's economic theory of value [44, 46].

Lindberg et al., [47] employed a choice experiments to assess residents' attitudes towards the costs and benefits of increased tourism on a Danish community. Hearne and Santos [48] assessed tourists' preferences towards measures designed to manage in a sustainable manner a protected area in Guatemala.

The DCE approach is based on a choice-based questionnaire that presents a number of choice sets comprising two or more alternative packages asking respondents to make trade-offs between newly developed attributes/levels and their respective acquisition cost. According to McLeod et al. [49] the key assumption informing the DC approach is "that the value of an option depends on the value of its attributes". The major limitation of the DCE approach lies in the necessity of restrict the number of attributes selected for analysis, which implies previous knowledge/identification of the most important attributes. The literature suggests the number of attributes analyzed in a DCE ranges from 2 to >10, "with a mean of 5" [49]. In terms of sample size, Orme [50] recommended sample sizes with a minimum of 200 respondents- Marshall et al. [51] reported an average of 259 in health care studies with 40% of the sample sizes in the range of 100 to 300 respondents [52, 53]. This case study focuses on a small number of key attributes and levels identified in the literature as it was deemed inappropriate to ask respondents to analyze a several attributes in each choice profile. Quite often "researchers must rely on a limited array of profiles in the final DCE" [54]. For example, with 5 attributes, 4 of them with 2 levels and a fifth one with 3 levels, if all possible combinations of attributes and levels were to be compared in a two-alternative design, then there would be 48 (2^4*3) different profiles and 1128 different combinations of two-alternative choice questions [54]. In this study we used a design with 8 choice sets with 2 choices profiles per set [55, 56].

The survey instrument in this study was designed to help respondents to answer the questions promptly and correctly, "in order to minimize the time spent by a respondent to complete it". As each choice comprises an alternative characterized by changes in one or more attributes and the status quo "current situation", with all attributes defined by its basic levels, the model is binary [47].

Usually, respondents are also invited to answer several questions about their socio-demographic background and to express their opinions. The questionnaire includes a few lines of text with information regarding the experiment and the different attributes and levels considered to better illustrate the process. A choice experiment is prone to be regarded as too complex and a very attention demanding task, which may be the case with respondents with low level of basic skills in terms of academic qualifications. "Choice models have a reputation for being difficult to interpret." [57]. However data available on the sociodemographic profile of visitors traveling to Madeira suggests that in light of the fact that we included only 5 attributes to be compared, we could be confident about respondents' ability to effectively appraise the alternative packages under analysis (**Table 2**).

The identification of possible market development demands, in an initial phase, a thoroughly analysis of local concerns and needs, to identify attributes for analysis in a second stage, based on the choice experiment. Local operators voiced their concerns over increased competition from the local lodgment sector. These operators suggested several improvements such as extra number of amenities to diversify the current offer with extended holidays in view. A certain number of ideas and worries were identified such as the high tax burden and competition from other sectors. Some operators stressed the need for "family activities" being available to the sector as a whole. Based on the opinions voiced by operators, a preliminary draft of survey instrument was developed based on the attributes deemed more import and levels chosen for empirical analysis and pretested among tourists. The full factorial experimental design included in total 24X3 combinations. The full factorial design

Total	Daily aver pp	Total pp	Perc.
Package	117,21€	980,85€	
Accommodation	43,57€	390,80€	35,27%
Air transport	33,11€	388,81€	26,80%
Restaurants, bars, coffee shop and discotheques	19,13€	185,60€	
Local/regional transport	6,18€	58,13€	5,00%
Supermarket and grocery	5,20€	52,65€	4,21%
Cultural, sportive and entertainment related activities	2,56€	23,80€	2,07%
Shoes, clothes and accessories	5,41€	56,95€	4,38%
Crockery, décor and embroidery	4,25€	40,43€	3,44%
Health care	1,58€	21,01€	1,28%
Other expenditures	2,56€	24,40€	2,07%

Source: Authors.

Table 2.
Detailed package Share of different types of expenditure per stay.

was reduced based on fractional experiment design computed via SPSS. In the end, eight different alternatives were compared with the current situation. As usual in the literature, the combination of attributes in each scenario, and the combination of "choice" scenarios forming each choice sets followed a number of criteria such as orthogonality and "level balance between attributes".

Choice experiments as a stated preference technique are employed to examine visitors' preferences *vis a vis* with several hypothetical market developments based either on the introduction of a range of new attributes or on upgrades of the current offer. Choice experiments also allows the estimation of visitor's willingness to pay for such developments. As mentioned above, the concept of choice experiment is based upon two key theoretical foundations, the Lancasterian consumer theory and random utility theory. Lancasterian theory postulates that utility derives from the attributes embedded in a particular product or service. Random utility theory theorizes that individual utility (U) comprises two components: a systematic or deterministic component (V) that can be measured and an unobserved or stochastic component (ε). The level of utility enjoyed by an individual in scenario A can then be expressed as follows:

$$U_{iA} = V_{iA} + \varepsilon_{iA}. \quad (1)$$

Moreover, the systematic component can be expressed as a linear function of explanatory variables. Therefore, the deterministic component can be defined as follows:

$$V_{iA} = \beta' x_{iA}. \quad (2)$$

The econometric analysis of the deterministic component is based on maximum likelihood estimation methods and on the assumption of independently and identically distributed (i.i.d.) Type 1 extreme value error terms, with a scale factor μ and a variance σ2, where μ > 0 and σ2 = π2/6 μ2. Based on the i.i.d. type I hypothesis it is possible to use the multinomial logit model.

5. Results

The outline of a DCE design requires attention to the identification of the most important attributes and the assessment of the possible levels. An experimental design was constructed by using the orthogonal option available in SPSS. We considered only 5 attributes to avoid the risk of over complexity and to ensure adequate "levels of length, layout, specific wording and comprehensibility" [49, 58].

The questionnaire included questions about respondents' socio-demographic background in terms of age, gender, income, professional background and marital status in order to test the influence of individual heterogeneity on stated preferences. They were asked to rate their degree of interest in extra services such as a meal service and to rate their impressions about the sector range of tourism facilities and experiences. Similarly, respondents were invited to rate, on a 1–5 Likert scale, the extent to which they think the sector offered price advantages. We defined the following attributes: farming and related activities; outdoor activities; cultural and learning activities; loyalty program. In relation to each of the eight different choices presented in the survey, respondents were invited to choose between two alternatives, one being the package of with upgrades in one or several attributes, but at a higher daily price, and the other being the current situation, in terms of the facilities provided at the current daily prices. As visitors preferences are expected to vary between individuals, owing to their own preferences and inclinations, as well as in different levels of income and age, we also tested the impact of a number of interactions between the main attributes and certain segments.

The questionnaire provided a certain amount of background information to illustrate the nature of the experiment. The econometric analysis was performed on STATA version 16. We first conducted the analysis based on the traditional McFadden's choice model, multinomial logit model, in order to determine the signal and magnitude of the coefficients β1, β2, β3, β4 and β5 in order to determine the relative importance of each of the attributes, based on the following equation:

$$V = \beta 0 + \beta 1 * \text{Farming} + \beta 2 * \text{Price} + \beta 3 * \text{Outdoor} + \beta 4 * \text{Promotion} \\ + \beta 5 * \text{Information} \quad (3)$$

However, to account for the presence of heterogeneity across respondents we also estimated a mixed-logit model that included random effects as well as fixed effects "because the random effects can often account for potential variation in relative preferences across participants [59]".

Regarding demographic characteristics of the respondents, approximately %51, of the respondents were male, Portuguese nationals (6%), German nationals and had at undergraduate level of education (91.4%). Approximately 46% of the sample is retired. Data concerning respondent's activities carried out while on the region suggest that less than 50% undertakes recreation activities such trekking/walking along the levadas (43%), visits to museums (28,4%), go to the beach (17,2%), visits to monuments (29,6%) and tours around the island (28,44%) and visits to touristic routes (31,3%) and boat trips (2.3%). Only tasting sessions organized around the local gastronomic attract a high number of visitors (64,%).

Only 300 completed questionnaires were eligible for the econometric analysis. The sample cannot be considered as representative of entire the RT segment as we included only rural establishments willing to participate in this research. Therefore, the results must be interpreted as merely indicative but deemed helpful to support operators access to accurate "data" on visitors' preferences.

In line with Louviere *et al.* [44], Hearne and Salinas [58] we estimated the McFadden's choice model, which is a specific case of the more general conditional logistic regression model. In this study, we focus our analysis of the relationship between the choice of an upgrade version of the current product booked by visitors and respondents' preferences in a number of key attributes as well as a number of respondents socio-demographics characteristics.

Farming, outdoor, are defined as discrete changes in the level of the attribute, meaning that the coefficients can be interpreted as reflecting the impact of discrete changes from one level to the next, as with dummy variables [58]. The significance and sign of the coefficients reported in **Table 3**, as well as the related results in **Table 4**, follow expectations. Each of the attribute coefficients has the expected sign and is highly significant. The coefficient of price is negative, which indicates that the probability of choosing the alternative scenario decreases as the prices increases. The negative coefficient of the variable "price" also reflects visitors' preferences for lower daily rates, which confirmed the fears expressed by a number of operators. The coefficients of the variables "farming", "outdoor", "promotions" and "information" are positive, indicating that respondents′ probability of choosing the alternative scenario increases with increased levels of attribute supply.

The results indicate that the sign for age and income is negative mean that both older and well-off visitors were less likely to be interested in alternative scenarios offering access to extra experiences as a result of "less need for experiences" or simply "resistance to change generally". Moreover, the higher the level of interest in "rest and relax", the higher the probability of supporting the status quo.

Moreover, the results imply that respondents had a significant preference for greater access to information, outdoor and farming and rural activities. As the base alternative refers to the business as usual scenario, the positive coefficient associated with the variable "family" indicates that if a respondent travels in family, such visitor is more likely to choose the alternative over the business as

Variable	Homogenous Specification		Interaction terms Specification	
	Coef. (St. Error)	Sig	Coef. (St. Error)	Sig.
Farming	0,518	0,001	1,004	0,027
Price	−0,147	0,000	−0,277	0,040
Outdoor	0,631	0,006	0,894	0,069
Information	0,663	0,001	1,048	0,021
Promotions	0,337	0,031	1,401	0,003
Price*Age	...	...	0,027	0,019
Farming*Age	...	...	−0,100	0,259
Outdoor*Age	...	...	−0,053	0,588
Informations*Age	...	...	−0,078	0,380
Promotion*Age	...	...	−0,220	0,019
Constant	−1,37946	0,000	−1,414	0,000
Wald chi2	117,65		129,47	
Prob	0,000		0,000	
Log	−586,46		578,501	

Source: Authors.

Table 3.
Results of the conditional logit (McFadden's) choice model.

	WTP	WTP	WTP/Av. Rate	
Farming	3,512	3,618	5,16%	5,32%
Outdoor	4,281	3,223	6,29%	4,74%
Information	4,496	3,777	6,61%	5,55%
Promotions	2,287	5,050	3,36%	7,42%
Price*Age	...	0,097	...	0,14%
Farming*Age	...	−0,639	...	−0,94%
Outdoor*Age	...	−0,191	...	−0,28%
Informations*Age	...	−0,282	...	−0,41%
Promotion*Age	...	−0,792	...	−1,16%

Source: Authors.

Table 4.
Willingness to pay measures (euros).

usual scenario. The impact of age is negative; therefore, younger people are more likely to support the alternative compared to older people, probably as a result of a more pro-activity stance. The relationship between income and probability of choosing the alternative scenario is not consistent, though respondents with the highest level of income are most likely to opt for the alternative scenario, which does not apply to the remaining levels of income. In fact, the relationship between income and the alternative is negative for levels 5 and 6 and not significant for the remaining levels. The expected probability of a respondent reporting the level 7 of income (the highest) is 0,365. The is, we expected 36,5% of the individuals with a level of income of 5000€ or more to choose the alternative. As income increases from 1 to 6, respondents are less likely to choose the alternative. The degree of interest in alternative scenarios decreases as the level of income goes up. It is worth to mention that increases in income from level 1 to 2, and so on, do not lead to significant different probabilities of choosing the alternative scenario. Therefore, increases in income from each level to, the immediately following one do not lead to significantly different probabilities of choosing the alternative scenario.

Only highly well-paid individuals are ready to consider upgrades of the current version of the product being supplied by the sector.The results also indicate that males are less likely to pick the alternative scenario than females (relative-risk ratio 0,769). Older visitors are the least interested in choosing an alternative scenario based on upgrades.

Table 4 indicates that the effect of reporting a age level of 2 instead of 1 is 0,02 decrease in the expected probability of choosing the alternative scenario. In a similar vein, reporting an age level of 7 instead of 6 leads to a decrease in the expected probability of 0,0176. Both effects are significant at 1% level of significance.

Model 2 estimates the impact of interactions among the attribute variables and other variables. The interaction price*age was significant, which suggests that the negative impact of price is partially offset by age. On the opposite direction, the interest on promotions decreases with age.

The values of the Willingness to Pay (WTP) suggests the visitors display a relatively high WTP for "information" and "promotions". The results displayed in **Table 4** indicate that visitors are ready to pay an extra of 4,496€ for information, which corresponds to 6,61% of the average daily rate. Another interesting observation corresponds to the significance that respondents ascribe to access to outdoor activities in the vicinity of the establishment. This is translated into a willingness to pay 4,28€. The access to higher levels of outdoor activities would increase respondents' level of utility by 4,28€. Regarding the provision of rural activities, visitors would be willing to pay 3,5€ per day for the introduction of such activities.

The results provided so far indicate that respondents value to a certain extent hypothetical development in the product being offered. Nevertheless, in this assessment, account must be taken, that from the point of view expressed by respondents, a large majority just prefer the status quo. In fact, the results indicate that 22,9% of the respondents will choose the alternative scenario, while 70,1% will choose the business as usual scenario.

The evidence from this study suggests that visitors prefer the status quo (in 70% of the cases), which suggests the operators must develop the sector in a way that does not compromise the integrity of the experience sought by respondents. In general, visitors understand the RT experience as an opportunity to relax and gaze. This is in contrast to how the sector and the literature are inclined to interpret visitors′ needs. Operators should direct their attention to developing a product based on increased information because visitors are prepared to pay to find out more about the region's offer and to allow greater choice. Developments in this regard must take into account that any changes must place the visitor's needs as the focal point of product development. For the time being visitors prefer the status quo even if they are ready to pay a slight increase in the daily rates to have access to textured layers of information. The discussion provided so far indicates that a consumer-oriented approach is paramount in this study as far development is concerned. Given the low WTP computed, the increased focus on upgrades should not be pursued at the expense of financial viability and comfort issues.

Tourists explicitly stated they prefer the status quo because increased levels of price would contribute to lower levels of utility and to decreased probability of choosing the alternative scenario. As German and French nationals, represent almost 50% of the sample, their preferences must be examined in detail, and changes planned accordingly.

The results provided so far could assist us in putting forward a few suggestions. Operators should be encouraged to develop cooperative strategies in order to promote the sector and to share resources and fixed costs by negotiating with other (tourism agencies, operators). Past studies highlight aspects such as lack of attractiveness [14]; in such circumstances, "simply providing accommodation facilities does not guarantee demand" [14], and thus operators are required to develop a "package must be sufficient to attract and keep tourists, offering suitable opportunities for spending [51]. Another key issue to bear in mind related the fact that the development of new services "may require a significant investment either beyond the means of the business owner or greater than justified by potential returns" [14, 26]. Issues of lack of capacity to meet obligations in terms of high standards in "guest-service values" and quality of products and services matching tourists' demands and expectations have been also reported in the literature [21]. In several cases, the owner does not master certain key skills in terms of marketing, innovation and design of product. More importantly, the process of renovating or converting old proprieties or developing from scratch new amenities/services may involve prohibitive costs. In addition, most businesses report revenues below average and low occupancy levels. The figures for 2019, occupancy vary between 25,1 per cent in January and 55,5 per cent in August.

In the relation to the daily prices rates, and for those tourists making a reservation at an establishment outside the capital city, we must add the cost of hiring a car, which imposes an increase of *% in relation to average daily rate. In a number of cases, rural establishments lack a sufficient number of facilities and amenities in their villages and surrounding areas, which forces visitors to rely on a rent a car to travel around, otherwise they had little choice but to carry out a very limited number of in-situ activities. Visitors interested in experiencing several activities such as are forced to travel to a number of specific locations.

As regard the price, account should be taken that around one third of the total cost's correspondent to price of air tickets. As a result of the high prices of the air tickets most respondents are facing budgetary constraints. Data available on the sector (as of 2016) shows that visitors spend on average 2,07% on cultural, sportive and entertainment related activities. Accommodation, transport and restaurants account for 77,55%. Data referring to the year 2015 and to tourism satellite account at regional level points into same direction. The share of the item "cultural, recreational, entertainment and other services" stands at 2,2% of the total amount of visitors´ level expenditure.

6. Conclusions

RT is expected to act as a key agent of local development in rural areas, in line with the regional development research. Contrary to the experience elsewhere, with the beginnings of RT taking place decades ago in close link with farm-tourism and homecoming visits, RT in Madeira is exclusively focused in offering accommodation since its inception and exhibits signs of decline after ten years in the market.

Based on the results of the choice experiment, it is shown that most visitors are interested in farming activities but over-value "price cuts" and price discounts (loyalty reward programs). Furthermore it is also shown that visitors would appreciate further improvements in terms of outdoor activities and information available. However, visitors are willing to pay, on average, very small amounts for further improvements. Consequently, investments in farming facilities are likely to be too costly, and a likely threat to the economic viability of the sector. In line with the evidence provided based on other geographical setting, the recovery of the initial investment highly unlikely. Therefore, further developments at the micro level should be preceded by sound economic analysis (in terms of cost–benefit) and realistic assessment of the sector´ potential to attract a high number of visitors and their and willingness to pay for improvements. However, even if we accept that firm are exposed to significant business risk (further investments involves a high risk of business failure), the evidence of slowing or stagnant growth suggests that further action is needed to reduce the over-dependence upon basic pricing strategies leading to exaggerated price cuts. Managers cannot expect to charge higher prices, in an attempt to recover its investment, but they may succeed in adding new features (based on outdoor activities and cultural events) to the standard product. The preference for extra information, which includes tips about indoors and outdoors activities, cultural events, gastronomy and facts about rural lifestyles suggests that visitors value "diversity" in terms of rural activities and are anxious to know better all the options available. In that sense, a high density of natural and attractions in rural areas, may offer an opportunity to increase the quality of the tourism experience and to minimize the current focus on price cuts.

The traditional financing mechanism based either in retained profits to invest or public funding are unlikely to generate sufficient resources leading to an appropriate level of investment in infrastructure, facilities and e-business models. "Funds" are only a small part of the problem. Policy makers, industry actors, need to understand the big picture. Agriculture is an issue of common interest to the public and private sector, with long term impacts on rural areas.

The analysis of RT as a catalyst for growth cannot be detached from discussion of the ′new′ regional policy philosophy, which emphasizes endogenous development, the intensification of the economic-social modernization processes taking place in the global arena and the affirmation of the neo-liberal paradigm. Due to the growing resistance to further financial transfers (From the European Union), it may eventually be impossible to justify further public investment. However, some peripheral regions may not ever reach minimum levels of economic and social viability, and any progress in terms of minimization of market inefficiencies, (a key argument of the neo-liberal orthodoxy against the current state of affairs on islands), optimal policy formulation and efficient application of all funds available will demand further "public investment". A sudden reduction in the amount of financial transfers in the middle of an on-going process of modernisation may stop the momentum in terms of economic and institutional modernisation. Further research is needed to understand tourism development in line with the new development paradigm faced by islands and further research is needed to estimate how much people are willing to pay for maintaining aesthetical pleasant landscapes.

In conclusion, RT is not performing the expected "developmental/regenerative" role [14], neither is a "counterpoint to mass, package-type tourism". The evidence provided so far, rejects the over-optimistic approach regarding RT [25] as the sector impacts only marginally upon the economic prosperity of rural areas. If successful, the RT sector may offer a rescue to the overall tourism industry in Madeira facing increasing pressure from cheaper and more dynamic destinations.

Operators should be encouraged to develop cooperative strategies in order to promote the sector and to share resources and fixed costs by negotiating with other (tourism agencies, operators).

Tourism development offers the opportunity to test new business models. This study offers an excellent opportunity to examine the current efforts to develop alternative market segments in a mature and consolidated destination. And this study also pledges for an integrated approach to understand RT development.

Author details

António Almeida[1] and Luiz Pinto Machado[2]*

1 Universidade da Madeira, Funchal, Portugal

2 Universidade da Madeira and CEFAGE-Center for Advanced Studies in Management and Economics

*Address all correspondence to: luizpintomachado@staff.uma.pt

References

[1] Perales, R. (2002). Rural tourism in Spain. Annals of Tourism Research, 29(4), 1101-1110

[2] Sharpley, R. (2003). Tourism, modernisation and development on the Isla nd of Cyprus. Challenges and policy responses. Journal of Sustainable Tourism, 11(2/3):246-265

[3] Liu, A. (2006). Tourism in rural areas: Kedah, Malaysia. Tourism Management, Vol. 27, pg. 878-889

[4] Sharpley, R., & Vass, A. (2006). Tourism, farming and diversification: An attitudinal study. Tourism Management, 27(5), 1040-1052.

[5] Fons, S., Fierro, M., Patino, G., (2011). Rural tourism: A sustainable alternative. Applied Energy, Vol. 88, pg. 551-557

[6] Berkel, D. Verburg, P. (2011). Sensitising rural policy: assessing spatial variation in rural development options for Europe. Land Use Policy, Vol. 28, pg. 447-459

[7] Pina, I., Delfa, M. (2005). Rural tourism demand by type of accommodation. Tourism Management, Vol. 26, pg. 951-959

[8] Albacete-Saez, C. A., Fuentes-Fuentes, M. M., & Llorens-Montes, F. J. (2007). Service quality measurement in rural accommodation. Annals of Tourism Research, 34(1), 45-65

[9] Andriotis, K. (2006), "Researching the development gapbetween the hinterland and the coast – evidence from the island of Crete", *Tourism Management*, Vol. 27, pp. 629-663.

[10] Zasada, I, Piorr, A. (2015). The role of local framework conditions for the adoption of Rural Development Policy: an example of diversification, tourism development and village renewal in Brandenburg, Germany. Ecological Indicators, Vol. 59, pg. 82-93,

[11] Bardolet, E. e Sheldon, P. (2008), "Tourism in archipelagos: Hawai'i and the Balearics", Annals of Tourism Research, Vol. 35, n.º 4, pp. 900-923.

[12] Briedenhann, J., Wickens, E. (2004). Tourism routes as a toll for the economic development of rural areas – vibrant hope or impossible dream?. Tourism Management, Vol. 25, pg. 71-79

[13] Burton, R. J. F. (2004). Seeing through the "good farmer's" eyes: towards developing an understanding of the social symbolic value of "productivist" behaviour. Sociologia Ruralis, Vol. 14, pg. 195-216

[14] Sharpley, R. (2002). Rural tourism and the challenge of tourism diversification: The case of Cyprus. Tourism Management, 23, 233-244

[15] Saxena, G., Ilbery, B. (2010). Developing integrated rural tourism: actor practices in the English/Welsh border. Journal of Rural Studies, 26: 260-271

[16] Cawley, M., Gillmor, D. (2008). Integrated rural tourism, Concepts and Practice. Annals of Tourism Research, 35(2): 316-327

[17] Lane B., Kastenholz E., Lima J., Majewsjki J. (2013). Industrial Heritage and Agri/Rural Tourism in Europe. Directorate General for Internal Policies

[18] Aytuğ, H., Mikaeili, M. (2017). Evaluation of Hopa's Rural Tourism Potential in the Context of European Union Tourism Policy. Procedia Environmental Sciences, Vol. 37, pg. 234-245.ISSN 1878-0296,, https://doi.org/10.1016/j.proenv.2017.03.039.

[19] Iorio, M., Corsale, A., 2010. Rural tourism and livelihood strategies in Romania. J. Rural Studies. 26, 152-162.

[20] Molera, L., & Albaladejo, I. (2007). Profiling segments of tourists in rural areas of South-Eastern Spain. Tourism Managment, 28, 757-767.

[21] Fleischer, A., Pizam, A., (1997)"Rural Tourism in Israel" Tourism Management, Vol. 18, No. 6, (1997), pp. 367-372.

[22] Lekakis, Stelios, Dragouni, Mina Heritage in the making: Rural heritage and its mnemeiosis at Naxos island, Greece,Journal of Rural Studies, Volume 77,2020,Pages 84-92, ISSN 0743-0167

[23] López-i-Gelats, F., Milán, M.J., Bartolomé, J., (2011). Is farming enough in mountain areas? Farm diversification in the Pyrenees. Land Use Pol., Vol. 28 (4), pg. 783-791

[24] Brandth, B. Haugen, M. (2011). Farm diversification into tourism – implications 837 for social identity? Journal of Rural Studies′, Vol. 27, pp. 35-44

[25] Barke, M. (2004). Rural tourism in Spain. International Journal of Tourism Research, 6: 137-149

[26] Fleischer, A., Felsenstein, D. (2000). Support for rural tourism. Does it make a difference?. Annals of Tourism Research, 27(4): 1007-1024

[27] Kumar, Suneel. (2020). Technology and innovation: Changing concept of rural tourism – A systematic review. Open Geosciences. 12. 737-752. 10.1515/geo-2020-0183.

[28] García-Falcón, J., and Medina-Munõz, D. (1999), 'Sustainable tourism development in islands, a case study of Gran Canaria', Business Strategy and the Environment, Vol 8, No 6, pp 336-357

[29] Rodríguez, J., Parra-Lopez, E., Yanes-Estévez, V. (2008). The sustainability of island destinations: tourism area life cycle and teleological perspectives. The case of Tenerife. Tourism Management, 29(1): 53-65

[30] Butler, R (2000). Issues and implications of tourism development in Maritime Regions, Perspectivas de Desenvolvimento para as Regiões Marítimas. Actas do VII Encontro Nacional da APDR, 15-28,

[31] Weaver, D. (1995). Alternative tourism in Montserrat. Tourism Management, Vol. 16(8), pg. 593-604.

[32] Hampton, M.P. and Christensen, J. (2007) 'Competing Industries in Islands: a New Tourism Approach.' Annals of Tourism Research, 34 (4) 998-1020.

[33] Mutlu D H, Atik M. 2014. Rural Tourism and Vernacular Architecture: An Example from Elmalı, Antalya. American International Journal of Contemporary Research. 2014; 4:69-72.

[34] Tao, T. H., & Wall, G. (2009). Tourism as a sustainable livelihood strategy.Tourism Management, 30,90-98

[35] Rodríguez-Pose, A., (2000). Economic convergence and regional development strategies in Spain: The case of Galicia and Navarre, EIB Papers, Vol. 5 (1), pg. 88-115

[36] Oppermann, M., (1996), Rural Tourism in Southern Germany, Annals of Tourism Research 23(1):86-102, DOI: 10.1016/0160-7383(95)00021-6

[37] Sharpley, R., Roberts, L. (2004). Rural Tourism-10 Years On. International Journal of Tourism Research, Vol. 6, pp. 119-124

[38] Hjalager, A. (1996). Agricultural diversification into tourism, evidence of a European Community development

programme. Tourism Management, 17(2), pg. 103-111.

[39] Albaladejo-Pina, I., Díaz-Delfa, M. (2009). Tourist preferences for rural house stays: Evidence from discrete choice modelling in Spain. Tourism Management, Vol. 30(6), pg. 805-811

[40] Walford, N. (2001). Patterns of development in tourist accommodation enterprises on farms in England and Wales. Applied Geography, 21(4), 331-345

[41] Frochot, I. (2005), "A benefit segmentation of tourists inmrural areas: a Scottish perspective", Tourism Management, Vol. 26, pp. 335-346

[42] Fleischer, A., Tchetchik, A. (2005). Does rural tourism benefit from agriculture? Tourism Management, 26, 493-501.

[43] Osti L, Cicero L. Tourists'perception of landscape attributesin rural tourism. Worldwide Hospital Tour Themes.2018;10(2):211-21

[44] Louviere et al., 2000 J.J. Louviere, D. Hensher, J.D. Swait Stated choice methods: analysis and application Cambridge University Press, Cambridge (2000)

[45] Hensher, D.A. Rose, J.M. Greene, W.H. (2005), Applied choice analysis: a primer, Cambridge University Press, Cambridge

[46] Seddighi, H. R., and A. L. Theocharous. 2002. A model of tourism destination choice: A theoretical and empirical analysis. Tourism Management, Vol. 23 (5), pg. 475-487

[47] Lindberg, K., Dellaert, B., G., C., and Rassing, C., R. (1999). Resident Tradeoffs: A ChoicemModeling Approach. Annals of Tourism Research 26: 3, 554-569.

[48] Hearne R., Santos, C., 2005. Tourists' and Locals' Preferences toward Ecotourism Development in the Maya Biosphere Reserve, Guatemala. Environment, Development, and Sustainability, Vol. 7, pg. 305-318

[49] McLeod C, Norman R, Schultz A, et al., (2019). Discrete choice experiment to evaluate preferences of patients with cystic fibrosis among alternative treatment-related health outcomes: a protocol. BMJ Open 2019;9:e030348. doi:10.1136/bmjopen-2019-030348

[50] Orme BK. Getting Started with Conjoint Analysis: Strategies for Product Design and Pricing Research. Madison, WI: Research Publishers LLC; 2006.

[51] Marshall D, Bridges J, Hauber AB, et al. Conjoint analysis applications in health - how are studies being designed and reported? An update on current practice in the published literature between 2005 and 2008. The Patient – Patient Centered Outcomes Research 2010;3:249 –56.

[52] Bridges J., Hauber A., Marshall D., Lloyd, A., Prosser, L., Regier, D., Johnson, F., Mauskopf, J., (2011). A checklist for conjoint analysis applications in health: report of the ISPOR Conjoint Analysis Good Research Practices Task Force. Value Health 2011, Vol. 14, pg. 403-13

[53] Lancsar E, Louviere J. Conducting discrete choice experiments to inform healthcare decision making: a user's guide. Pharmacoeconomics 2008;26:661-77.

[54] Foreman, A., Friedel, J., Hayashi, Y., Wirth, O., (2021), Texting while driving: A discrete choice experiment, Accident Analysis and Prevention 149 (2021) 105823

[55] Johnson F., Lancsar E., Marshall D., Kilambi V., Muhlbacher A., Regier D.,

Bresnahan B., Kanninen B, Bridges J., (2013). Constructing experimental designs for discrete-choice experiments: report of the ISPOR Conjoint Analysis Experimental Design Good Research Practices Task Force. Value Health, Vol. 16, pg. 3-13.

[56] Street, D.J., Burgess, L., Viney, R., Louviere, J., 2008. Designing discrete choice experiments for health care. In: Ryan, M., Gerard, K., Amaya-Amaya, M. (Eds.), Using Discrete Choice Experiments to Value Health and Health Care. Springer, Dordrecht, pp. 47-72.

[57] STATA (2019), STATA choice models reference manual, StataCorp. 2019. Stata: Release 16. Statistical Software. College Station, TX: StataCorp LLC.

[58] Hearne, R. e Salinas, Z. (2002), "The use of choice experiments in the analysis of tourism preferences for ecotourism development in Costa Rica", Journal of Environmental Management, Vol. 65, pp. 153-163.

[59] Hauber A., González J., Groothuis-Oudshoorn C., Prior T., Marshall D., Cunningham C., IJzerman M., Bridges J. F. (2016). Statistical methods for the analysis of discrete choice experiments: A report of the ISPOR conjoint analysis good research practices task force. Value in Health, Vol. 19, pg. 300-315.

Chapter 7

An Assessment of the Impact of the Tourism Sector on Regional Economic Development in Gauteng Province, South Africa

Daniel F. Meyer

Abstract

South Africa is facing three main developmental problems, including high levels of poverty, unemployment, and inequality. The tourism sector allows for a relatively easy entry into the local market for small businesses and entrepreneurs and has the potential to create jobs and subsequently, income. Tourism development could be utilised as a driver for economic growth and development. The main objective of this research was to assess the impact of the tourism sector on economic growth and development in South Africa, focusing on the Gauteng Province which, is the economic hub of the country and even Africa. The methodology utilised was based on a quantitative design, using secondary time series pooled panel data approach including, all the municipal entities in the region. Annual data from 2000 to 2019 were used to analyse the impact of tourism on economic growth and development. Tourism variables include measurements such as tourism spending and international tourism trips. Results confirm the tourism-growth nexus and the sector allows ease of market entry for small businesses, resulting in employment creation and income for the poor in developing regions if promoted via effective policy implementation, even in regions where tourism is not the leading sector.

Keywords: economic development, economic impact, Gauteng region, South Africa, tourism

1. Introduction

Tourism is an important economic sector and contributes 10.4% of global GDP; one in ten jobs are provided via tourism; and 1.4 billion international trips are completed per annum [1]. Globally nearly 60 percent of the population lives in poverty [2], and this number is expected to increase as a result of COVID-19. South Africa has some of the highest levels of poverty, unemployment, and even income inequality. These factors are often obstacles to sustainable economic growth. Within the context of this study, tourism could be defined as all types of travel activities and affects local communities through the environment, employment, and social systems [3]. People living in poverty have limited choices due to a lack of knowledge, opportunities, and skills and find it difficult to exit the poverty trap [4].

Within this context of growing unemployment and poverty, tourism as an economic sector is widely accepted as a sector that could help marginalised and disadvantaged regions achieve higher growth levels with more employment opportunities [5]. As an economic sector, tourism is one of the leading sectors and a fast-growing sector [6]. However, COVID-19 has had a massive impact on tourism on a global scale. As an industry, tourism acts as an export industry, attracting foreign income into the receiving region and is, therefore, the main source of income for many poor developing regions [7]. Traditional economic sectors such as mining and manufacturing have in recent decades struggled to contribute significantly to poverty reduction and employment, and tourism has proven to be able to play that role in developing regions [8]. Samini, Sadeghi, and Sadeghi [9] go as far as to state that tourism could act as a driver or engine of growth as the sector creates employment opportunities, which could lead to more disposable income and increases in government tax revenues [9, 10]. Growth in the sector will result in the diversification of regional and local economies, which is especially important for regions that have focused economies dominated by single resource sectors. These economies are vulnerable to economic and other shocks [11]. The diversification of such economies could minimise the impact of shocks. According to Pedrana [11], the tourism sector could drive investment, allow for employment opportunities, facilitate the inflow of money, and contribute to infrastructure development.

Despite acknowledging the importance of tourism for economic growth and development, the section is still ignored as an important economic sector due to lack of data and reporting [12, 13]. Phiri [14] states that due to limited data and subsequent limited quantitative academic research, the exact impact of tourism as an economic sector on economic growth and development, especially in developing regions and nations, are limited. In the South African context, it is needed to determine the impact of tourism activities on growth and development [15]. For this reason, the focus of this study is on the economic hub of South Africa and even Africa, the Gauteng province, which is situated in central South Africa and includes cities such as Johannesburg, Pretoria, and Soweto. The region is the only major metropolitan region in the world not located next to an ocean or major river system. This study's contribution is found in the qualitative and quantitative analysis of the dynamic impact of tourism on regional economies using panel data in a developing region.

2. Literature review

This section of the study consists of three parts: an explanation of concepts and definitions, a theoretical foundation, and results from other empirical studies. In terms of concepts and definitions, tourism is defined as people travelling from one region or place to another for more than one night for reasons including holiday, recreational and sports, cultural, medical, and business activities [11]. According to [1] tourism is a multidimensional concept and linked to many economic activities and consists of individuals, communities, businesses, and organisations as well as places that collaborate to provide a tourist experience. Tourism could facilitate employment opportunities for all and even for people with lower levels of education [5].

As this study investigates the links between the tourism sector and regional economic development, the concept of regional economic development is also defined. Stimson, Stough, and Roberts [16] provide comprehensive definitions of regional economic development supported by other researchers. They state that regional economic development is a similar concept if compared to economic development, but on a regional scale. They list the following components of the concepts: regional economic development is a process where regional stakeholders including business, communities, and local government attempt to stimulate economic opportunities through the involvement of all economic sectors, employment and quality of life. According to Stimson et al. [16] the regional economic development process includes policy formulation, planning, analysis, strategy development, and resource application; while regional economic development product includes employment, wealth creation, investment, infrastructure development, and quality of life.

This section contains empirical results on the relationship between the tourism sector and regional economic development. Rosentraub and Joo [17], over three decades, assessed several metropolitan regions across the world where investments have been made to attract larger numbers of tourists and human capital. Key results from this research indicate that sports and amusement-related tourism activities allow for most gains and benefits for regional development. Some authors [18] analysed the linkages between the tourism industry and economic development processes. The study focused on regional economic development, investigating the importance of tourism geography's economics and the increasing significance of networks and new information and communication technologies. Tourism complexity and economic development's role is essential, and the importance of the term 'glocal' in tourism development processes.

From a European perspective, authors such as [19] assessed tourism's contribution to regional economic resilience in Europe. The results confirmed the significance of tourism as a factor of regional economic resilience. Furthermore [20] assessed domestic and international tourism's impact on the regional economic growth for 179 European regions from 1999 to 2009 using GDP per capita as the dependent variables with other variables such as human capital, technological capital, institutional and social features of the regions. The final results established both domestic and international tourism do positively affect economic growth. Dana, Gurau, and Lasch [21] analysed the potential of two regions regarding tourism in rural areas as a source of entrepreneurship opportunities supporting regional economic development in rural, regional France. According to Alberti and Giusti [22], tourism plays a major role in regions' economic development and competitiveness. This paper assesses the clustering of tourism and cultural heritage activities that could lead to regional competitiveness in the Motor Valley cluster in Italy from 1999 to 2011. This region is globally known for its motorsport industry, with firms involved in this industry since 1800 in facilities, institutions, cultural heritage, museums. All role players collaborate, leading to competitiveness in regional tourism. The study results confirmed the importance of establishing clusters focused on tourism interlinked with a cultural heritage for regional competitiveness. Cortes-Jimenez [23] analysed the role of tourism on regional economic growth in regions of Spain and Italy using panel data econometric methods. The results indicate that tourism through both domestic and international tourism contribute positively to regional economic growth. However, it should be noted that the impacts and patterns of impacts on regional development differ from region to region.

Vieira and Santos [24] examined the role of tourism for regional economic development in Portugal in terms of spatial interrelations between municipal regions. Results indicate significant spatial interrelations between and across the various municipalities and that spatial tourism clustering occurs in coastal locations. Besides, tourism was confirmed via an econometric analysis as a significant driver of regional economic development. Petrevska and Manasieva Gerasimova [25] analysed the impact of tourism on regional economic development in the south-western section of Macedonia, focusing on tourism flows, the capacity of tourism accommodation, and tourism spending from 2003 to 2010. Findings from the analysis are that this region is a leading region in Macedonia but still needs to achieve its potential. This situation exists due to limited policy formulation and implementation. Surugiu and Surugiu [26] studied the relationship between the tourism sector and economic growth in Romania from 1988 to 2009 using econometric cointegration Granger causality methods, vector error correction model (VECM), and impulse response functions. The findings show that tourism expansion does granger cause economic growth. Results from this study place a focus on the requirement for effective tourism development strategies. Fundeanu [27] analysed the role of tourism clusters in the south-west Oltenia region, looking at tourism potential, tourism diversity, strengths, and weaknesses of the region. The study found that tourism clusters are catalysts for regional development, and the competitive advantages of such clusters should be the focus of public policies and strategies.

Gunderson and Ng [28] studied the impact of tourism on regional development in the rural USA. Tourist spending could result in increased demand for regional goods and services, eventually leading to employment creation and an increase in disposable income. The results indicate that public policy effectiveness, sustainable natural resource management, and community development could allow for tourism development and regional development. Tourism positively affects regional economic performance. Klytchnikova and Dorosh [29] analysed the role of tourism on regional economic development in Panama's poor regions using a social accounting matrix model. The paper used the impact of tourism spending on growth and poverty at the regional level. The results indicate that tourism has a large impact on the regional economy and is also an important multiplier in the local economy. The sector also allows for important benefits to the poor. Mishra, Rout and Mohapatra [30] considered tourism an important sector to promote regional economic growth and analysed the import of India's sector from 1978 to 2009 using econometric methods. Time-series econometric models were used for the analysis from 1978 to 2009. The results indicate the existence of unidirectional causality running from tourism activities to economic growth.

Wen-li [31] analysed the impact of tourism on economic growth in regional China since the 1990s. The study results indicate a significant impact of tourism on regional economic growth and allow for diversified and balanced development. He and Zheng [32] analysed the Sichuan region from 1990 to 2009 in China and the tourism sector's impact with abundant resources on the regional economy. Over the last decade, the contribution of tourism to the provincial GDP has been increasing annually. Results indicated that a bi-directional relationship exists between tourism development and economic growth. Yang, Fik, and Altschuler [33] analysed tourism-related economic multipliers from regional input–output tables for 30 Chinese provinces looking at tourism variables, including income, employment, and employment multipliers. Interesting findings reveal that the output and employment multipliers of tourism are positively associated with regional economic development.

Rogerson [34] states that uneven development is a reality of South Africa's spatial economy's structure with leading and lagging regions. Tourism has been identified as a vital economic sector for regional development. This paper assesses the 23 distressed regions in dire need of economic development. These regions rely mostly on domestic tourism, and local natural assets should be maximised with effective policy implementation. Meyer and Meyer [8] conducted a study using regional tourism statistical data from 2001 to 2013 for two geographical areas in a developing region in South Africa. The results indicate that tourism in these regions has a significant impact on economic growth as the sector does include low skilled workers in a labour-intensive industry and allows for a range of benefits for regions that include employment and income.

Lastly, the tourism sector could also have negative impacts, especially on the environment and sustainable development. Effective policies should be in place to allow for strategies to prevent the environment's deterioration [35]. Pedrana [11] believes that tourists could negatively affect local cultures. These negative impacts could be minimised with local community involvement and if local partnerships and cooperation exist between the private and public sectors. The tourism industry could play an important role in protecting the social and cultural integrity of a community [36].

3. Methodology

The research methodology is based on a quantitative analysis using both descriptive and econometric methods to achieve the primary objective. In this study, the tourism sector's impact on a developing region is analysed using secondary data from Global Insight [37]. Annual data from 1996 to 2019 were used and analysed firstly utilising trends and correlation analysis and secondly by using a pooled econometric panel approach including the five municipal areas (see **Figure 1** for details) in the Gauteng province in South Africa. Gauteng province comprises of the following municipal regions:

- The city of Johannesburg, Metropolitan region (COJ)
- City of Tshwane, Metropolitan region (COT)
- Ekurhuleni, Metropolitan region (EKR)
- Sedibeng district municipal region (SDM)
- West Rand district municipal region (WRDM)

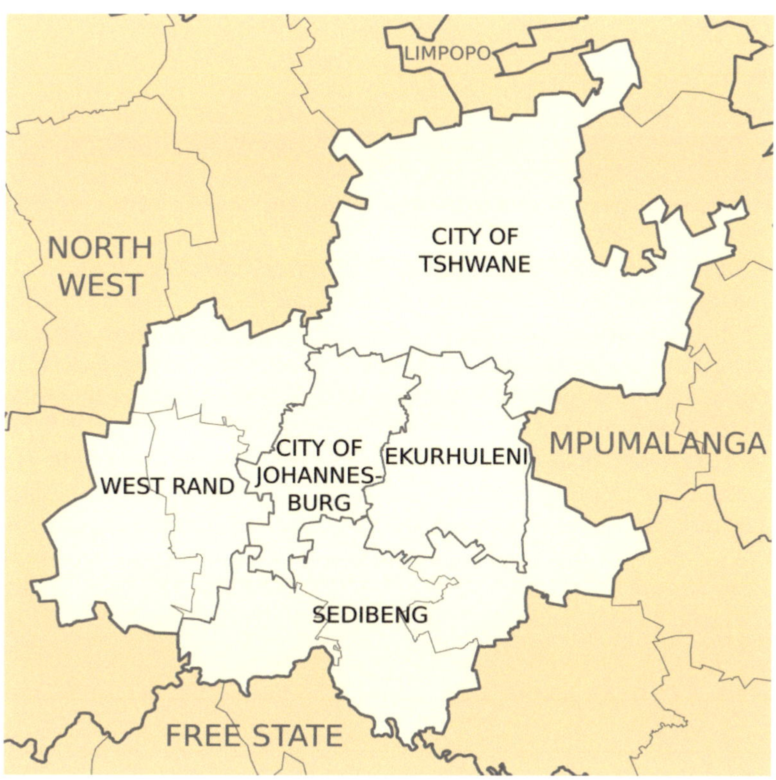

Figure 1.
Gauteng municipal regions. Source: https://en.wikipedia.org/wiki/List_of_municipalities_in_Gauteng

The variables used in the panel econometric analysis consisted of GDP per capita representing economic development and growth as the dependent variable, with the following independent variables: gross value added (GVA) in the tourism sector; the number of jobs in the tourism sector; spending per capita in the tourism sector; and the number of international trips to the region. The panel data for the Gauteng region were analysed using a multiple regression. A multiple regression includes many variables to predict changes in the dependent variable [38]. All of the variables were converted into natural logarithms. The variables were set as follows with the abbreviations as used:

Dependent variable (Y) = Economic development and growth (GDPC - the log format LGDPC)

Independent variables (X):

- Gross value added in the tourism sector (GVAT – the log format LGVAT)
- Number of jobs in the tourism sector (JOBST – the log format LJOBST)
- Spending per capita in the tourism sector (SPENDT – the log format LSPENDT)
- Number of international tourist trips (INTTT – the log format LINTTT)

Pedroni [39] formulated Eq. (1) represents the basic model for the panel data analysis:

$$Y_{it} = a_i + \delta_i t + \beta_i X_{it} + e_{it} \tag{1}$$

Where:

Y_{it} = Dependent variable.

a_i = Intercept term.

δ_i = Parameter that, together with a_i allows the individual linear trends and individual effects to be observed respectively [6].

β = k × 1 vector of parameters that were estimated based on the explanatory variables.

X_{it} = 1 × k vector of observations of the explanatory variables, t = 1,. .., T; i = 1,. . .

In this econometric analysis process, several models were used to test for long and short-run relationships between variables: (1) unit root tests to determine the level of stationarity of the variables and model selection; (2) long-run relationships between the variables using either a panel ARDL of Fisher-Johansen cointegration test leading to regression analysis using FMOLS and DOLS equations; (3) Granger causality test to assess causality between all the variables; (4) and model stability diagnostic tests. To simplify the analysis of results from all tests and to place all variables on the same scale, all variables were converted into the natural logarithm format. A panel data set was created for the five municipal regions within the Gauteng Province with 120 observations. Eq. (2) represents the basic equation for pooled panel data econometric models, as proposed by Brooks [40]:

$$y_{it} = \alpha + \beta x_{it} + u_{it} \ and \ I = 1\ldots,N; t = 1,\ldots,T \tag{2}$$

Where y_{it} is the dependent variable, α is the intercept term, β is a $k \times 1$ vector of parameters to be estimated on the explanatory variables, and x_{it} is a $1 \times k$ vector of observations on the explanatory variables, t = 1,.., T; i = 1, i denotes countries in the panel and t denotes time dimension. The i subscript denotes the cross-section and t the time series. The model from the function described in Eq. (1) can be listed as follows in Eq. (3):

$$LGDPC_t = \propto_1 + \sum_{j=1}^{k} \beta_{1j} LGDPC_{t-j} + \sum_{j=1}^{k} \lambda_{1j} LGVAT_{t-j} + LJOBST_{t-j} + LSPENDT_{t-j} + LINTTT_{t-j} + u_{1t} \quad (3)$$

Where α_n is the constant, β_n, λ_n are the coefficients, K is the number of lags, and u_{1t} and u_{2t} are the stochastic error terms, which are also known as shocks in the model. The following tests were used to determine the stationarity level for all the variables: Im, Pesaran and Shin W-stat test; the ADF - Fisher Chi-square test; and the PP - Fisher Chi-square test. Model selection was based on the unit root results. Stationarity of all variables at levels or I(0), a panel VAR analysis would be estimated. In contrast, if all variables were stationary at 1st difference or I(1), the Fisher Johansen panel cointegration test for long-run relationships should be estimated. Lastly, if a mixture of variables were presented, the optimal option would be a panel ARDL method as an estimation.

4. Results and discussion

4.1 Descriptive analysis

As mentioned earlier, the focus area selected for this study is the Gauteng Province in South Africa. This region was chosen due to the following reasons. Firstly, it is the economic hub in South Africa and the African continent; secondly, it is rich in cultural and historical places and events. The region is the largest metropolitan region in the world not located adjacent to the ocean or a major water body. Based on the aforementioned information, this research study's main objective is to assess the dynamic economic impact of tourism on the Gauteng province, using both descriptive and time-series approaches.

In terms of the descriptive analysis, eight key variables were selected as indicated in **Table 1**, to compare the five municipal regions with the total Gauteng province. Variables were analysed regarding growth rates, trends, and contributions to the study region. The different descriptive variables are analysed individually. Firstly, the GDP at constant prices is analysed. The Gauteng province had strong economic growth from 2009 to 2014 of 3.1% per annum, but growth has been low and slow from 2014 to 2019 at 1.1%. The COJ contributed the most to the provincial GDP of 44%, followed by the COT. The two peripheral regions of SDM and WRDM only contributed 4% and 3.8% to the province. The COT had the highest annual growth rate of 1.3% from 2014 to 2019. Secondly, Gauteng has a declining situation regarding GDP per capita with a negative growth rate of −1.2% from 2014 to 2019. Of the five municipal areas, both COJ and COT had much higher GDP per capita values and negative growth rates. SDM had the lowest GDP per capita at less than half of the two leading regions, namely COJ and COT. Thirdly, in terms of population density, Gauteng province had a density of 649 per sqkm and has an increased rate of 2.8% per annum. COJ and EKR metros had the highest densities of 2630 and 1562, with high levels of increases. The two more rural regions of SDM and WRDM have much lower at 208 and 176,

Variables	Year	COJ	COT	EKR	SDM	WRDM	Total Gauteng
GDP at constant prices (R million)	2009	412238	250160	176621	38031	45544	922596
	2014	465263 (2.6)	302464 (4.2)	208049 (3.6)	43682 (2.9)	43870 (−0.7)	1065330 (3.1)
	2019	493223 (1.2)	321164 (1.3)	215681 (0.7)	44852 (0.5)	42004 (−0.9)	1116927 (1.1)
GDP per capita	2009	101171	92190	59140	42686	57163	80489
	2014	97276 (−0.8)	95823 (0.8)	61371 (0.8)	45353 (1.3)	52188 (−1.7)	80965 (0.1)
	2019	90851 (−1.3)	90325 (−1.2)	57141 (−1.4)	43132 (−0.9)	47211 (−1.9)	76040 (−1.2)
Population density (People per sqkm)	2009	1871	321	1178	174	153	483
	2014	2270 (4.3)	386 (4.1)	1375 (3.3)	190 (1.8)	164 (1.4)	569 (3.6)
	2019	2630 (3.2)	445 (3.1)	1562 (2.7)	208 (1.9)	176 (1.5)	649 (2.8)
Jobs in the tourism sector	2009	48465	26388	27212	4415	7815	114295
	2014	55497 (2.9)	30903 (3.4)	32730 (4.1)	6098 (7.6)	7984 (0.43)	133213 (3.3)
	2019	60436 (1.8)	37722 (4.4)	38591 (3.6)	6298 (0.7)	9117 (2.8)	132155 (2.8)
Disposable income (R million)	2009	217674	146803	128552	29777	30830	553637
	2014	238950 (1.9)	170929 (3.3)	146191 (2.7)	35471 (3.8)	33702 (1.9)	625246 (2.6)
	2019	256716 (1.5)	185860 (1.8)	157971 (1.6)	39085 (2.1)	36034 (1.4)	675669 (1.6)
International tourism trips as a ratio to total trips	2009	0.22	0.20	0.21	0.20	0.25	0.21
	2014	0.26 (3.6)	0.24 (4.0)	0.27 (5.7)	0.25 (5.0)	0.32 (5.6)	0.25 (3.8)
	2019	0.33 (5.4)	0.30 (5.0)	0.34 (5.2)	0.32 (5.6)	0.39 (4.4)	0.32 (5.6)
Tourism spending per capita	2009	5613	5043	3818	1769	2324	4483
	2014	7278 (5.9)	7022 (7.9)	5385 (8.2)	2657 (10.1)	3559 (10.6)	6151 (7.5)
	2019	7356 (0.2)	7165 (0.4)	5730 (1.3)	2717 (0.45)	3986 (2.4)	6360 (0.7)
Tourism GDP per capita	2009	5934	5374	4030	1857	2513	5376
	2014	5898 (−0.1)	5681 (1.2)	4307 (1.4)	2131 (2.9)	2628 (0.9)	5198 (−0.7)
	2019	4708 (−4.0)	4527 (−4.1)	3583 (−3.4)	1687 (−4.2)	2210 (−3.2)	4558 (−2.5)

Source: [37]. Note: () brackets contain annual growth percentages between observations.

Table 1.
Descriptive analysis of municipal regions in Gauteng Province.

respectively. Fourthly, in terms of tourism jobs, Gauteng had 132 155 people working in tourism with a growth rate of 2.8% per annum. Compared with the five sub-regions, COJ contributes most with 39% of total jobs in the province, followed by EKR with 25% and COT at 24%. COT and EKR had the highest growth rates in jobs per annum of 4.4% and 3.6%, respectively.

Regarding disposable income, the Gauteng province has been increasing at 1.6% since 2014 in line with the slow growth of GDP growth. COJ again contributes most regarding the income of 38% of the total provincial income, followed by COT at 27%. SDM had the highest increase income of 2.1% coming from a low base, while the increase in income for the rest of the municipal regions had low growth rates of between 1.4–1.8%. Next, the international tourist trips are analysed as a ratio of total tourism trips to the regions. The WRDM had the highest ratio of 0.39, followed by EKR with a ratio of 0.34. All of the regions had relatively high growth rates of above 4.4% per annum. Tourism spending per capita for Gauteng province showed high growth levels from 2009 to 2014 at 7.4%, but as with the rest of the economy did the tourism sector also show much lower spending at 0.7% per annum since 2014 2019. COJ and COT had higher tourism spending per capita than the provincial values, with low growth rates of 0.21% and 0.4%, respectively. The three metro regions had much higher tourism spending levels if compared to the two more rural regions. By far, the WRDM had the highest increase in tourism spending per capita of 2.4% per annum since 2014. Lastly, the GDP per capita analysis indicates negative growth rates of – 2.5% per annum from 2014 to 2019, while only COJ has slightly higher tourism spending per capita than the province. As with the provincial growth rates, all of the municipal regions also had negative growth rates of −3.18% to −4.17%.

Table 2 indicates the correlation coefficients for the variables included in the econometric analysis. GDP per capita has a positive and significant relationship with all other variables, with GVA in the tourism sector with the highest coefficient of 0.93, followed by disposable income at 0.83. The two variables with the highest shared

Variables	GDP per capita	Disposable income	Pop density	GVA in tourism sector	Jobs tourism sector	Int tourism trips	Spending in tourism per capita
GDP per capita	1.0000						
Disposable income	0.8314	1.0000					
Pop density	0.5214	0.8374	1.0000				
GVA in tourism sector	0.9291	0.9414	0.7064	1.0000			
Jobs in tourism sector	0.7318	0.9506	0.8754	0.8630	1.0000		
Int tourism trips	0.7943	0.9329	0.8362	0.9150	0.9733	1.0000	
Spending in tourism per capita	0.7017	0.8609	0.6326	0.7965	0.9020	0.9112	1.0000

Table 2.
Correlation coefficients analysis.

coefficient are jobs in the tourism sector and international trips of 0.97, followed by jobs in the tourism sector and disposable income with a coefficient of 0.95.

4.2 Econometric analysis

Table 3 illustrates the results obtained from the unit root testing. Unit root tests are the first step before cointegration estimations are completed. The reason for this important step is to prevent the use of non-stationary variables, as this may produce spurious results [41]. Unit root tests also assist in the selection of the final long-run estimation model. This analysis results reveal that all variables are non-stationary at levels I(0), while all variables become stationary at 1st difference. Based on the unit root test results, it could be concluded that the Fisher - Johansen panel cointegration test should be utilised to assess the long-run relationships between the variables as all variables are stationary at the same level or at 1st difference.

The next step in the process was to determine the direct nexus between the independent variables and GDP per capita, which was used as a proxy for economic development. The Fisher - Johansen panel cointegration test was utilised as this test is most suitable when all variables are of order I(1). The results are presented in **Table 4**. The Fisher - Johansen cointegration test is used to test for long-run cointegration. [6] define cointegration as the systematic co-movement between variables in the long-run. The results indicate that the Trace and Max-Eigen tests provide evidence of a cointegrating relationship between the variables, at a 1 percent significance level. It could be concluded that the results obtained from the Fisher - Johansen cointegration test confirm a long-run equilibrium relationship between tourism and economic growth. The subsequent step is to determine the exact impact of tourism on economic growth in the study region.

The Fisher - Johansen test indicates a long-run relationship between the variables. This relationship needs to be confirmed, and the strength of the various relationships between the variables need to be established via a regression analysis with coefficients. For this purpose, the two types of estimation methods utilised are the Fully Modified Ordinary Least Squares (FMOLS) and the Dynamic Ordinary Least Squares (DOLS) models. A consideration of various forms of residual-based panel method results indicate that these models generally out-perform single-equation estimation techniques [39]. Firstly, the results of both the FMOLS and DOLS models are listed in **Table 5**. With LGDPC as the dependent variable, the FMOLS model results indicate that all of the independent variables are significant and positive predictors of GDP per capita (LGDPC) at a significance level of 1%,

Variables	Levin, Lin and Chu test		Im, Pesaran and Shin W-stat		ADF - Fisher Chi-square test	
	I (0)	I (1)	I (0)	I (1)	I (0)	I (1)
LGDPC	0.1036	0.0001*	0.1515	0.0001*	0.1534	0.0002*
LGVAT	0.1394	0.0008*	0.7478	0.0001*	0.8529	0.0003*
LJOBST	0.0305**	0.0001*	0.7798	0.0003*	0.8116	0.0001*
LSPENDT	0.9484	0.0045*	0.9478	0.0175**	0.9723	0.0074*
LINTTT	0.0005*	0.0427**	0.3174	0.01311**	0.5237	0.0031*

Null hypothesis: Unit root.
**indicates 1% statistically significant.*
***indicates 5% statistically significant.*
Source: own elaboration.

Table 3.
Panel unit root tests: P-values.

Hypothesised No. of CE(s)	Fisher Stat. (from trace test)	Probability	Fisher Stat. (from max-eigen test)	Probability
None	185.6	0.0001*	121.40	0.0004*
At most 1	93.04	0.0003*	44.46	0.0007*
At most 2	57.05	0.0005*	49.02	0.0000*
At most 3	19.89	0.0303	14.47	0.1527

**indicates variables are statistically significant at 1% and ** at 5%.*

Table 4.
The Fisher - Johansen panel cointegration test.

Method	Variables	Coefficient	Std. Error	t-statistic	Prob.
FMOLS	LGVAT	0.8796	0.0854	10.2886	0.0001*
	LJOBST	0.2328	0.0535	4.3463	0.0003*
	LSPENDT	0.4473	0.0862	5.1881	0.0002*
	LINTTT	0.0993	0.0805	1.2328	0.2212
DOLS	LGVAT	0.8609	0.1315	6.5432	0.0001*
	LJOBST	0.2963	0.1014	2.9219	0.0139*
	LSPENDT	0.1740	0.1589	1.0952	0.2968
	LINTTT	0.4196	0.0999	4.2001	0.0015*

**indicates variables are statistically significant at 1% and ** at 5%.*
Dependent variable: LGDPC.
Independent variable: LGVAT; LJOBST; LSPENDT; LINTTT.

Table 5.
FMOLS and DOLS results.

except for international tourist trips (LINTTT). GVA in tourism (LGVAT) has the highest coefficient of 0.88, meaning that a 1% increase in LGVAT could increase by an increase of 0.88% in GDP per capita. Spending on tourism (LSPENDT) has the second-highest coefficient of 0.45, followed by the number of jobs in tourism (LJOBST) with a coefficient of 0.23.

Similar results have been estimated using the DOLS method. The only major difference between the two methods is that only spending on tourism (LSPENDT) is not a significant predictor of GDP per capita for the DOLS model. At the same time, international tourist trips are, however, a significant predictor. Similar coefficients were estimated for both models. This analysis's results are that all the variables could be accepted as significant predictors of GDP per capita. Paci and Marrocu [20] found similar results in that both domestic and international tourism positively influences regional economic growth. Alberti and Giusti [22] also found a positive relationship but added that if all regional role players collaborate, it could lead to more competitiveness in regional tourism.

Table 6 presents the results of the pairwise Granger-causality tests indicating short-run relationships. The purpose of the Granger causality test is to determine which variable causes changes to any of the other variables in the model [42]. The focus of the analysis is on the dependent variable, namely GDP per capita. The results indicate that changes in GVA in tourism impact GDP per capita, while GDP per capita does cause changes in both jobs in tourism and spending in tourism. No causality was detected between GDP per capita and international tourist trips. Causality between other variables excluding the official dependent variable allows

Null Hypothesis:	Obs	F-Statistic	Prob.
LGVAT does not Granger Cause LGDPC	110	7.4393	0.0010*
LGDPC does not Granger Cause LGVAT		0.1007	0.9043
LJOBST does not Granger Cause LGDPC	110	0.5872	0.5577
LGDPC does not Granger Cause LJOBST		7.4212	0.0010*
LSPENDT does not Granger Cause LGDPC	85	0.8280	0.4406
LGDPC does not Granger Cause LSPENDT		10.4710	9.E-05*
LINTTT does not Granger Cause LGDPC	85	0.8497	0.4313
LGDPC does not Granger Cause LINTTT		1.9399	0.1504
LJOBST does not Granger Cause LGVAT	110	0.2084	0.8122
LGVAT does not Granger Cause LJOBST		14.7989	2.E-06*
LSPENDT does not Granger Cause LGVAT	85	7.2153	0.0013*
LGVAT does not Granger Cause LSPENDT		11.0183	6.E-05*
LINTTT does not Granger Cause LGVAT	85	7.3938	0.0011*
LGVAT does not Granger Cause LINTTT		5.4569	0.0060*
LSPENDT does not Granger Cause LJOBST	85	4.6972	0.0118*
LJOBST does not Granger Cause LSPENDT		6.7106	0.0020*
LINTTT does not Granger Cause LJOBST	85	14.4359	4.E-06*
LJOBST does not Granger Cause LINTTT		7.1791	0.0014*
LINTTT does not Granger Cause LSPENDT	85	11.1663	5.E-05*
LSPENDT does not Granger Cause LINTTT		10.7987	7.E-05*

**Rejection of null hypothesis at 5% significance level.*

Table 6.
Granger causality tests.

for interesting results. Changes or increasing GVA in the tourism sector cause changes in tourism jobs and not vice versa, so new value-adding products and services, in this case, cause more jobs in the sector. Also, bi-directional causality relationships exist between spending in the tourism sector and GVA in tourism; between international tourism trips and GVA in tourism; spending and jobs in the tourism sector; international tourist trips and tourism; and between international tourist trips and spending in tourism. Mishra et al. [30] found similar results whereby tourism activities Granger-cause changes in regional economic growth.

Lastly, the econometric model is tested in terms of stability using residual diagnostics. In addition to the various aforementioned statistical procedures, diagnostic statistics were used to determine whether the residuals were distributed normally. Three residual diagnostic tests were performed, namely the Jarque-Bera normality test, the serial correlation test, and a heteroscedasticity test. To achieve this, the histogram of the residuals device was used. The histogram of residuals and the Jarque-Bera statistic shows that the data are normally distributed, and the results gained are valid. In terms of the serial correlation, both tests had AC values above 0.5, suggesting no autocorrelation between the variables. The results further suggested that there was no conditional heteroscedasticity among the variable.

5. Conclusion and recommendations

This study's primary objective was to assess and evaluate the impact of the tourism sector on the regional economy of the Gauteng Province in South Africa. Research on the impact of tourism on regional economies is relatively limited if compared to other economic sectors. One of the reasons for this situation is that data on the tourism sector is limited as it is not counted as a formal economic sector. Only sections of the industry are counted in detail, such as the hospitality industry. The tourism sector cuts across many formal economic sectors making the quantification thereof difficult. Data on a provincial or regional level are even more limited than on a national level. The objectives of this study were achieved via both descriptive and econometric data analysis.

The study's main results indicate that the tourism sector has a significant positive impact on economic growth in the Gauteng provincial region. This study's outcome is important as it reduces the uncertainty surrounding tourism and its impact on regional economies. Results indicate that tourism could contribute significantly to economic growth per capita in the study region. A 1% increase in the tourism sector's gross value-added activities could lead to between a 0.86 to 0.89 percent increase in GDP per capita. This indicates that tourism does have the potential to decrease unemployment and further contribute to alleviating poverty and improving living standards for people in local regions.

For tourism to make even greater contributions to regional economies, close cooperation is required between key regional role players such as the business community via business chambers, provincial and local government, and local communities. An effective regional tourism organisation that has as its goals as sustainable tourism development is important. It could help promote the region, initiate new regional projects, share information, and improve coordination among industry leaders. Also, cooperation between the public and private sectors could ensure natural environment protection, leading to a more attractive and marketable region.

As with most research studies, this paper also had a few limitations. The findings of this study are based upon the results in the Gauteng region, which consists of a range of municipal regions. These sub-regions differ in terms of the level of development but forms a coherent entity. This region is a leading economic region in South Africa but not in tourism as it is located in-land. Results from this region may differ from results of coastal regions such as the Western Cape or Kwa-Zulu Natal regions. Data sets per municipal area were also only available only from 2000 up to 2019 but provided sufficient data for the analysis. However, the listed limitations allow for future research studies such as the comparison of in-land and coastal regions or regions with established tourism sectors versus regions where tourism has not been developed.

This study indicates that the tourism sector could even play a critical economic developmental role in regions that are not primarily focused on the tourism sector. For tourism to be a regional economic driver, a relatively clean environment is a requirement, as well as a diversified tourist product offering or tourism sector complexity. This relates to sustainable traditional economic sectors such as the mining and manufacturing sectors in collaboration with the tourism sector. South Africa and its regions and provinces have unlimited potential within the tourism sector due to its history, cultural diversity, rich biodiversity, and natural beauty. The tourism sector should be the main industry to revitalise the ailing economy.

Author details

Daniel F. Meyer
College of Business and Economics, University of Johannesburg, South Africa

*Address all correspondence to: dfmeyer@uj.ac.za

References

[1] Cooper C. Essentials of tourism. SAGE Publications Limited; 2020.

[2] Todaro MP, Smith SC. Economic development. 12th ed. Edinburgh Gate: Pearson Education Ltd; 2015.

[3] Harrill R. Residents' attitudes toward tourism development: A literature review with implications for tourism planning. Journal of Planning Literature. 2004;18(3), 251-266. DOI: 10.1177/0885412203260306.

[4] Ashley C, De Brine P, Wilde H. The role of the tourism sector in expanding economic opportunity. Corporate Social Responsibility Initiative Report. Cambridge, MA: Harvard University; 2007.

[5] Butler G, Rogerson CM. Inclusive local tourism development in South Africa: Evidence from Dullstroom. Local Economy. 2016;31(1), 264-281. DOI: 10.1177/0269094215623732.

[6] Seghir GM, Mostéfa B, Abbes SM, Zakarya GY. Tourism spending-economic growth causality in 49 countries: A dynamic panel data approach. Procedia Economics and Finance. 2015;23, 1613-1623. DOI: 10.1016/S2212-5671(15)00402-5.

[7] UNWTO (The United Nations World Tourism Organization). Why tourism? 2016. Available from: http://www2.unwto.org/content/why-tourism. [Accesses: 2020-04-12].

[8] Meyer DF, Meyer N. The role and impact of tourism on local economic development: A comparative study. African Journal for Physical, Health Education, Recreation and Dance. 2015;21(1:1), 197-214.

[9] Samini AJ, Sadeghi S, Sadeghi S. Tourism and economic growth in developing countries: P-VAR approach. Middle East Journal of Scientific Research. 2011;10(1), 28-32.

[10] Zuo B, Huang S. Revisiting the tourism-led economic growth hypothesis: The case of China. Journal of Travel Research. 2017;1-13. DOI: 10.1177/0047287516686725.

[11] Pedrana M. Local economic development policies and tourism: An approach to sustainability and culture. Regional Science Inquiry Journal. 2013;5(1), 91-99.

[12] Croes R, Vanegas M. Co-integration and causality between tourism and poverty reduction. Journal of Travel Research. 2008;47(1), 94-103. DOI: 10.1177/0047287507312429.

[13] Richardson RB. The contribution of tourism to economic growth and food security. Michigan, East Lansing: Michigan State University; 2010.

[14] Phiri A. Tourism and Economic Growth in South Africa: Evidence from linear and nonlinear cointegration frameworks. Managing Global Transitions. 2016;14(1), 31-53.

[15] Hanekom D. Local government tourism conference, Kempton Park, 30-31 March. 2015. Available from: https://www.tourism.gov.za/CurrentProjects/Past Projects/Local%20Government%20Tourism%20Conference%202015/Pages/Local-Government-Tourism-Conference-2015.aspx. [Accessed: 2020-10-27].

[16] Stimson RJ, Stough RR, Roberts BH. Regional economic development: analysis and planning strategy. Springer Science & Business Media; 2006.

[17] Rosentraub MS, Joo M. Tourism and economic development: Which investments produce gains for regions? Tourism Management.

2009;30(5), 759-770. DOI: 10.1016/j.tourman.2008.11.014

[18] Milne S, Ateljevic I. Tourism, economic development and the global-local nexus: Theory embracing complexity. Tourism Geographies 2001;3(4), 369-393.

[19] Bellini N, Grillo F, Lazzeri G, Pasquinelli C. Tourism and regional economic resilience from a policy perspective: Lessons from smart specialisation strategies in Europe. European Planning Studies. 2017;25(1), 140-153. DOI: 10.1080/09654313.2016.1273323

[20] Paci R, Marrocu E. Tourism and regional growth in Europe. Papers in Regional Science. 2014;93, S25-S50. DOI: 10.1111/pirs.12085.

[21] Dana LP, Gurau C, Lasch F. Entrepreneurship, tourism and regional development: a tale of two villages. Entrepreneurship & Regional Development. 2014;26(3-4), 357-374. DOI: 10.1080/08985626.2014.918182.

[22] Alberti FG, Giusti JD. Cultural heritage, tourism and regional competitiveness: The Motor Valley cluster. City, Culture and Society. 2012;3(4), 261-273. DOI: 10.1016/j.ccs.2012.11.003

[23] Cortes-Jimenez I. Which type of tourism matters to the regional economic growth? The cases of Spain and Italy. International Journal of Tourism Research. 2008;10(2), 127-139. DOI: 10.1002/jtr.646

[24] Vieira AC, Santos LD. Tourism and regional development: a spatial econometric model for Portugal at municipal level. FEP Working Papers 589, Universidade do Porto, Faculdade de Economia do Porto; 2017.

[25] Petrevska B, Manasieva Gerasimova V. Tourism in Regional Development: Empirical Evidence. Innovative Issues and Approaches in Social Sciences. 2012;5(2), 6-20.

[26] Surugiu C, Surugiu MR. Is the tourism sector supportive of economic growth? Empirical evidence on Romanian tourism. Tourism Economics. 2013;19(1), 115-132.

[27] Fundeanu DD. Innovative regional cluster, model of tourism development. Procedia Economics and Finance. 2015;23, 744-749.

[28] Gunderson RJ, Ng PT. Analysing the effects of amenities, quality of life attributes and tourism on regional economic performance using regression quantiles. Journal of Regional Analysis and Policy. 2005;35(1), 1-22. DOI: 10.22004/ag.econ.132298.

[29] Klytchnikova I, Dorosh P. Tourism sector in Panama: Regional economic impacts and the potential to benefit the poor. Natural Resources Forum. 2013;37(2), 70-79. DOI: 10.1111/1477-8947.12019.

[30] Mishra PK, Rout HB, Mohapatra SS. Causality between tourism and economic growth: Empirical evidence from India. European Journal of Social Sciences. 2011;18(4), 518-527.

[31] Wen-li ZHOU. A review on impact of tourism on economic growth. Economic Geography. 2011;8, 1-12.

[32] He LH, Zheng XG. Empirical analysis on the relationship between tourism development and economic growth in Sichuan. Journal of Agricultural Science. 2011;3(1), 212-223.

[33] Yang Y, Fik TJ, Altschuler B. Explaining regional economic multipliers of tourism: does cross-regional heterogeneity exist? Asia Pacific Journal of Tourism Research. 2018;23(1), 15-23. DOI: 10.1080/10941665.2017.1394335

[34] Rogerson CM. Tourism and regional development: The case of South Africa's distressed areas. Development Southern Africa. 2015;32(3), 277-291. DOI: 10.1080/0376835X.2015.1010713

[35] Cresco S, Querini G. The role of tourism in sustainable economic development. European Regional Science Association. 2003;1(1), 1-13.

[36] WTTC (World Travel and Tourism Council). Understanding the critical issues for the future of travel and tourism, March 2017. London: WTTC; 2017.

[37] Global Insight. Data analysis. 2020. Available from: https://g-insight.org/what/#data-analysis [Accessed: 2020-05-20].

[38] Gujarati DN, Porter DC. Essentials of econometrics. 4th ed. New York: McGraw-Hill Education; 2010.

[39] Pedroni P. Panel cointegration: asymptotic and finite sample properties of fooled time series tests with an application to the PPP hypothesis. Economic Theory. 2004;20, 597-625.

[40] Brooks C. Introductory Econometrics for Finance. Cambridge publishers; 2014.

[41] Ogbokor C.A. Foreign trade and economic growth in Namibia: A time series analysis. Vanderbijlaprk, NWU. (Thesis – PhD); 2015.

[42] Rivera MA. The synergies between human development, economic growth, and tourism within a developing country: An empirical model for Ecuador. Journal of Destination Marketing & Management. 2017;6(3), 221-232. DOI: 10.1016/j.jdmm.2016.04.002.